Nursery Care of Nonhuman Primates

ADVANCES IN PRIMATOLOGY

THE PRIMATE BRAIN
Edited by Charles R. Noback and William Montagna

MOLECULAR ANTHROPOLOGY: Genes and Proteins
in the Evolutionary Ascent of the Primates
Edited by Morris Goodman and Richard E. Tashian

SENSORY SYSTEMS OF PRIMATES
Edited by Charles R. Noback

NURSERY CARE OF NONHUMAN PRIMATES
Edited by Gerald C. Ruppenthal

Nursery Care of Nonhuman Primates

Edited by

Gerald C. Ruppenthal

University of Washington
Seattle, Washington

Associate Editor:
Dorothy J. Reese
University of Washington
Seattle, Washington

PLENUM PRESS · NEW YORK AND LONDON

Library of Congress Cataloging in Publication Data

Symposium on the Nursery Care of Nonhuman Primates, Battelle-Seattle Conference
 Center, 1977.
 Nursery care of nonhuman primates.

 (Advances in primatology)
 "Proceedings of a Symposium on the Nursery Care of Nonhuman Primates, held at
 the [Battelle-Seattle Conference Center] University of Washington, Seattle . . . May
 1—4, 1977."
 Includes index.
 1. Primates as laboratory animals — Congresses. 2. Primates — Nursery care — Congresses. I. Ruppenthal, Gerald C. II. Reese, Dorothy J. III. Title. IV. Series.
 SF407.P7S97   636'.98   78-32018
 ISBN 0-306-40150-9

Based on the Proceedings of a Symposium on the Nursery Care of Nonhuman Primates
held at the University of Washington, Seattle, Washington, May 1—4, 1977

©1979 Plenum Press, New York
A Division of Plenum Publishing Corporation
227 West 17th Street, New York, N.Y. 10011

Printed in the United States of America

Macaca nemestrina (pigtail macaque). Courtesy of the Child Development and Mental Retardation Center at the University of Washington.

PREFACE

The Infant Primate Research Laboratory at the University of Washington was conceived in 1970 as a small research unit primarily for support of two individual's interests in early development of nonhuman primates. Because of their research emphasis, a modest nursery was required to support a small population of animals for specific experimental studies. The laboratory experienced rapid growth when others at the University became interested in the use of monkeys as models for early development and mental retardation in humans.

In 1972 the unit was formally established as a core facility of the Child Development and Mental Retardation Center and the Regional Primate Research Center. This joint administrative and financial support allowed us to invest considerable effort in the development of normative data for rearing animals in our nursery as well as for identifying, documenting, and rearing subjects at high risk for neonatal death. As part of that effort, every attempt has been made to promote a multidisciplinary approach to questions associated with rearing nonhuman primates. This volume includes much of the information thus gathered. I feel that such an approach is essential to the promotion of scientific principles in rearing and has allowed the laboratory to contribute to primatology.

The chapters included here were presented at a symposium held May 1-4, 1977, at the Battelle-Seattle Conference Center, located near the University of Washington campus. Although the Center grounds are located within a residential neighborhood of Seattle, its spacious and forested surroundings provided a cloistered setting for the symposium. Presentation and discussion of the material in this four-day conference was, I am sure for all of us, a rewarding though exhausting experience. I would sincerely like to congratulate and thank all participants for their interest and efforts, and would hope that everyone found the interaction informative and stimulating.

Much of the success of the symposium was due to Ms. Tali Ott, Administrative Assistant to the Associative Directors of the Child Development and Mental Retardation Center, and symposium coordinator. Her tireless and enthusiastic attention to the myriad of problems connected with organization before and during the meeting were much needed and appreciated. The symposium would not have been possible without the joint sponsorship of the Child Development and Mental Retardation Center and the Regional Primate Research Center. I am extremely indebted to the directors of those centers, Dr. Irvin Emanuel and Dr. Orville A. Smith, for their interest and support in underwriting this effort. Also appreciated are the efforts of Mr. Henry Schulte, Administrator of the Child Development and Mental Retardation Center, who patiently answered my many questions and offered administrative advice and emotional support when it was badly needed. Special thanks are due to Ms. Kathleen Schmitt, Research Publications Editor at the Regional Primate Research Center, for her diligence and commentary during manuscript preparation and to Mrs. Ardith Greeny for her invaluable assistance in the production of this volume.

Finally, I would like to express my thanks to my friend, Gene P. (Jim) Sackett. Jim has been the driving force behind the development of the Infant Primate Research Laboratory, and he is much in evidence in this volume. His multi-faceted approach to the field of primatology and his enthusiastic investigative spirit have been instrumental in promoting our laboratory as a research resource for other investigators interested in studying nonhuman primates as models for human development and disease.

INTRODUCTION

Nursery care of nonhuman primates has evolved over the past 25 years from individuals taking an occasional problem animal home "in their pockets" to a rapidly growing research concern. There are many reasons for this growth. Of primary importance is that research in a number of biomedical and behavioral disciplines has emphasized the use of nonhuman primates as models for problems in the human population. Because of this, established research colonies have some percentage of animals that require nursery care because of research demands or because of clinical-husbandry problems. In addition, many laboratories are establishing breeding facilities, and with them, nurseries, to insure survival and useability of a maximum number of animals. Increasing restrictions on the importance of various species require the promotion of self-sustaining breeding colonies. In addition, colony managers are faced with increasing research demands for biologically 'clean' animals with known lineage. Zoological gardens are interested in saving the maximum numbers of primates that are reproductively active and can be attractively displayed. The list of reasons goes on, but taken together, all add to the requirement for-- and depend on the success of--nursery facilities throughout the world.

At first glance, the title of this volume might seem to be a misnomer, as much of the information included does not provide a "cookbook" approach to nursery rearing. Rather, it is meant to serve as a source of basic parameters and information about a number of questions and issues. These range from physiological, behavioral, and developmental normal values, to information concerning innovative support techniques for common laboratory species and for the more exotic species increasingly used in research and zoo display.

To that end, many chapters contain a blend of descriptive, theoretical, experimental, and normative data that have influenced both quantity and quality of nursery facilities. Some of these data have been gathered as part of normal rearing procedures but provide useful information for experimental applications. The converse is equally true. Much of the data are from experimental studies that can be used advantageously in nursery management and care.

This book is divided into five sections. Section I is devoted to the prenatal environment. It includes information concerning management of pregnancy including manipulations to enhance survival, clinical indications for Caesarean section, effects of stress on pregnancy outcome, characteristics of the placental environment, serological incompatibility, and prenatal diet deficiencies and their effect on postnatal development. Manipulations during pregnancy and termination of pregnancy, either for experimental purposes or from clinical necessity, can and do have important considerations for research and husbandry. Many of the offspring from pregnancies discussed in this section need nursery care.

Section II addresses early assessment techniques. Because there is great interest in skeletal maturation as it applies to both clinical and research applications, three chapters, giving differing approaches and results, are included. Data concerning bilirubin levels in neonates, useful as a husbandry tool and as a pediatrics model, are also found in this section. Finally, data on the effects of premature and Caesarean deliveries in the establishment and development of diurnal cyclicity are included in the last chapter of the section.

Section III is concerned with nursery care and management of three commonly used genera including <u>Papio</u>, <u>Saimiri</u>, and <u>Macaca</u>. Also included is a chapter concerning ponderal weight gains for mother-reared versus nursery-reared <u>M</u>. <u>nemestrina</u>. Two chapters are devoted to the design and use of apparatus for the care and assessment of high-risk animals, especially those at risk for respiration difficulties. The last chapter in the section is concerned with mortality and pathology in one nursery facility and points to problems encountered in the management of premature neonates.

Section IV deals with housing and early social development. It would seem unwise not to consider the effects of housing conditions on behavior, particularly as social behavior influences the survival rate of infants as well as adults. The first chapter in the section discusses this issue. Rearing animals in a nursery setting imposes at least a somewhat abnormal environment on them. The effects on intellectual capacity of differing degrees of isolation are assessed in the second chapter. The last chapter in this section is concerned with management and experimentation and their effects on behavioral development, physiology, reproduction, and maternal care.

Section V on exotic species deals with several less-known species of nonhuman primates currently being reared in research laboratories and in zoos. Basic parameters of several species and information concerning growth, development, and determinants of age are included.

It must be emphasized that the data presented here are not all-inclusive or exhaustive. Considerable data on nursery care remain to be analyzed. The information in this volume concerns only a small percentage of the many species of nonhuman primates that exist in the world today. Some of these species have been successfully reared--although possibly in small numbers--in various nursery facilities but data concerning their care and development are not available. Generalizations across species can be made to some degree but investigators must be aware that comparisons, even for closely related species, must be made guardedly. Comparisons within species are also incomplete at this time. Critical analyses of many issues can be made only by using data within nurseries and pooling data across nursery facilities. This requires an objective, unemotional, and multidisciplinary approach which will enhance our basic knowledge about nonhuman primates and facilitate optimal care.

My personal hope is that the information included in this volume will not only provide a reference for primatologists in established research and breeding facilities, but will also be a valuable resource for individuals who are currently faced with a "pocket full of monkey" and require nursery rearing for their nonhuman primates.

CONTENTS

SECTION I. PRENATAL INFLUENCES

SECTION II. EARLY ASSESSMENT

Chapter   Page

SECTION III. HEALTH, DIET, AND GROWTH

Chapter   Page

I. PRENATAL INFLUENCES

CHAPTER 1

CLINICAL INDICATIONS FOR CAESAREAN SECTION IN THE RHESUS MONKEY (Macaca mulatta)

C. J. Mahoney,* S. Eisele, and M. Capriolo

Regional Primate Research Center and
Psychology Department Primate Laboratory
University of Wisconsin
Madison, Wisconsin

*Present address: Laboratory for Experimental
    Medicine and Surgery in Primates (LEMSIP)
New York University School of Medicine
New York, New York

INTRODUCTION

Numerous reports indicate a high incidence of perinatal fetal mortality, ranging from 10 to 14% annually, in colonies of singly caged nonhuman primates (Lapin and Yakovleva, 1963; Hendrickx, 1966; Valerio, Courtney, Miller, and Pallotta, 1968; van Wagenen, 1972). A major portion of fetal deaths may be related to husbandry techniques. The frequency of both placental anomalies (Myers, 1971) and fetal dystocia during labor (Lapin and Yakovleva, 1963) has been attributed to a lack of exercise, constant stress from human interference, and possibly to dietary deficiencies.

In man, improved techniques of diagnosis are partly responsible for declining rates in fetal perinatal death. Furthermore, periodic examinations and a cognizance of previous obstetrical history permit a careful monitoring of the mother and fetus during gestation and enhance early detection of impending difficulties. By and large, such health programs are not practiced in breeding colonies of nonhuman primates. Understanding the causes of fetal demise in nonhuman primates would be valuable not only for increasing breeding efficiency but also for selecting animals as potential models for comparative studies in man.

3

Prompted by the high rate of fetal wastage in our colony of Macaca mulatta (Mahoney and Eisele, 1978), we began an ongoing study to 1) establish criteria of fetal and maternal well-being, 2) determine the causes of fetal death, particularly in late pregnancy, and 3) develop diagnostic procedures for the early detection of impending obstetrical problems. This chapter discusses the clinical indications for Caesarean section (C-section) and presents alternative approaches to the management of obstetrical complications in monkeys.

MATERIALS AND METHODS

The majority of female rhesus monkeys were singly caged during pregnancy. A total of 287 pregnancies occurred in 136 females in the nonexperimental breeding colony. Sixteen fetuses were born dead beyond day 139 of gestation (stillbirths). C-section, carried out for clinical reasons at or after expected term, resulted in delivery of six stillborn infants and 20 viable infants.

In the experimental breeding colony, the majority of late gestational fetal deaths occurred as a direct consequence of experimentation. However, the causes of five stillbirths were not related to the type of experiment and will be considered along with the nonexperimental group. Similarly included will be three viable C-sections performed in experimental pregnancies for clinical rather than experimental reasons.

Prepartum Examinations

Examinations, carried out at intervals of 1-5 days, were begun on day 150 of gestation or earlier for animals with a history of poor pregnancy outcome and in cases of suspected obstetrical complications. For routine examinations, unanesthetized females were manually restrained. However, for certain procedures, such as abdominal radiography or amniocentesis, animals were tranquilized by an intramuscular injection of ketamine hydrochloride (Vetalar, Parke-Davis) administered at a dosage of 10-15 mg/kg body weight.

The main objectives of the prepartum examinations were to 1) determine fetal viability, 2) predict the time of parturition, 3) determine presentation and position of the fetus, and 4) test for in utero fetal distress.

Determination of Fetal Viability. Viability could be readily determined by detecting fetal heart sounds or blood flow in the fetal placenta via an ultrasonic blood flow monitor (Model BF4A, Medsonics, Inc., Mountain View, CA). Details of the ultrasonic stethoscope and method of use have been described elsewhere (Mahoney and Eisele, 1976).

Prediction of Parturition. Previous experience indicated that the period of greatest risk for the fetus is during labor. Because of this, estimating the time of parturition was one of the major objectives of the antenatal examinations. In macaques, parturition can be predicted most reliably by monitoring the changes in physical characteristics of the cervix uteri (Mahoney, 1975; Mahoney and Eisele, 1978). As determined by rectal palpation, the cervix progressively softens until, by the last 1-10 days of pregnancy, it becomes completely nonpalpable. Vaginoscopic examination of the cervix reveals a concomitant progression in effacement of the portio vaginalis and dilatation of the external and internal ora.

Fetal Presentation and Position. Presentation and position of the fetus within the uterus were determined by rectal and transabdominal palpation. Occasionally, abdominal radiography was used to visualize compound malpresentations or, as in one case, the exact positioning of a set of twins. Because rhesus monkeys frequently assume a somewhat bipedal, rather than a quadripedal, position during second stage labor, fetal presentation and position will be described by the terminology used in human obstetrics. Anterior refers to the midline of the maternal abdomen; posterior relates to the maternal spinal column.

Evaluation of In Utero Fetal Distress. In human beings, passage of meconium into the amniotic fluid, especially when the fetus is in vertex presentation, strongly indicates a state of fetal hypoxia (Walker, 1954; Bernstine, 1960; Vorherr, 1975). Evidence of meconium staining in the rhesus monkey was obtained either by visualizing the chorioamniotic membranes through the dilated cervix or by amniocentesis.

Caesarean Section

Successful delivery of fetuses already suffering varying degrees of in utero hypoxia depended on minimizing stress to the mother and on rapid surgical procedure.

Anesthesia was obtained by intramuscular administration of ketamine hydrochloride (15-20 mg/kg body weight). To minimize the depressive effects of anesthesia on the fetus during preparation of the mother for surgery, only half of the calculated dose of ketamine was given initially. Administration of the remainder was delayed until 5 min before the anticipated time of incising the abdomen. When necessary, anesthesia was continued, after removal of the fetus, by a mixture of halothane (1%) and oxygen (2 1/min) delivered through an intratracheal tube.

Upon delivery through a transverse incision in the uterine fundus, the fetus was held at a level below that of the mother. This promoted drainage of blood from the placenta into the fetus. At the same time, gentle suction was applied to the nasopharynx to remove mucus and amniotic fluid. Clamping and severance of the umbilical cord were delayed for at least 60 sec after delivery of the fetus. Placental membranes were completely removed by external massage of the uterus. Following closure of the uterine incision, pitocin (0.5 units) was given intravenously at intervals of 5 min to promote uterine contractions, thereby reducing intrauterine hemorrhage.

RESULTS

Table 1 summarizes the causes of fetal death in late gestation in 22 nonexperimental and five experimental pregnancies (vaginal and Caesarean deliveries). Apart from postmortem evidence of anoxia, no specific cause of death could be established for eight fetuses.

TABLE 1.   CAUSES OF LATE GESTATIONAL FETAL DEATH (LESS THAN 139 DAYS) IN 22 NONEXPERIMENTAL AND 5 EXPERIMENTAL PREGNANCIES IN M. mulatta - VAGINAL AND CAESAREAN DELIVERIES

Number of Pregnancies	Causes of Death
8	Unknown causes
1	Placenta abruptio (complete)
1	Placenta praevia (central)
4	Placental insufficiency
1	Umbilical cord prolapse
1	Umbilical cord strangulation
7	Breech delivery/dystocia
2	Fruntum dystocia (in posterior occiput)
2	Maternal infection/anemia

Table 2 lists the clinical reasons for carrying out Caesarean delivery of viable infants in 20 nonexperimental and three experimental pregnancies.

Table 3 presents presumptive and definitive indications for C-section in nonhuman primates based on abnormal clinical signs and conditions in the mother or fetus. It should be emphasized that some of the signs listed do not constitute specific pathological entities in themselves. They may coexist with or occur as a result of other conditions. Furthermore, while some conditions present no immediate danger to the mother or her fetus, they may portend future complications.

Uterine Hemorrhage

The occurrence of uterine (vaginal) hemorrhage at any stage of pregnancy beyond the period of implantation (placental) bleeding is cause for immediate investigation.

Sudden profuse bleeding, especially in late pregnancy, is one of the few situations in the obstetrical management of nonhuman primates that requires urgent attention. Since not only the fetus but also the mother may be at risk, time should not be wasted in attempting to establish an exact diagnosis. Rather, C-section should be carried out immediately while attention is given to correcting maternal blood loss and electrolyte imbalance.

For practical purposes, differential diagnosis of the causes of bleeding can be divided into five categories: 1) normal antepartum hemorrhage, 2) placenta abruptio, 3) placenta praevia, 4) uterine rupture and 5) a number of miscellaneous conditions whose causes may not be immediately apparent (e.g., rupture of the placental marginal sinus, intraplacental hemorrhage, uterine and cervical pathologies).

Antepartum hemorrhage. Slight to moderate uterine bleeding was observed during the last 12-36 hrs before parturition in less than 5% of viable deliveries. The innocuous nature of this prepartum hemorrhage could be readily confirmed by vaginoscopic examination of the placental membranes through the dilated cervix.

Placenta Abruptio. Complete abruptio of the placenta, accompanied by severe hemorrhage of acute onset, occurred in two (0.70%) of the nonexperimental pregnancies exceeding 139 days duration. In one case, the fetus and later the mother died. In the other case, both survived following emergency C-section. Partial

TABLE 2.   CLINICAL REASONS FOR CAESAREAN DELIVERY
            OF VIABLE FETUSES IN 20 NONEXPERIMENTAL
            AND 3 EXPERIMENTAL PREGNANCIES IN M.
            mulatta

Number of Pregnancies	Clinical Reasons
6	Unnecessary: incorrect diagnosis of fetal death in 3, breech presentations with no attempt at manual version in 2, fetal tachycardia considered ominous sign of distress in 1
3	Placenta abruptio (1 complete, 2 partial)
4	Placenta praevia (2 central, 2 marginal)
7	Placental insufficiency with fetal distress
1	Pathological contraction ring
1	Breech presentation (incorrectable) of descended member of twins
1	Severe maternal vaginal prolapse

TABLE 3.   PRESUMPTIVE AND DEFINITIVE INDICATIONS FOR
            CAESAREAN SECTION IN THE RHESUS MONKEY
            BASED ON CLINICAL SIGNS AND CONDITIONS

1.   Uterine hemorrhage

2.   Prolonged gestation (more than 180 days)

3.   Protracted or recurrent unfruitful (dysfunctional) labor

4.   Fetal malpresentation/malposition/cephalopelvic disproportion

5.   Umbilical cord prolapse/strangulation

6.   Symptoms of in utero fetal distress, e.g., meconium passage, bradycardia, tachycardia

7.   Fetal death in utero

8.   Maternal hernias and prolapses

abruptio, involving only the secondary placental disc, occurred in two experimental pregnancies. Both were characterized by recurring episodes of moderate uterine hemorrhage during the last half of gestation. The mothers and the fetuses were monitored at frequent intervals until the latter had attained physiological maturity (approximately 145 days gestation), at which time C-section resulted in delivery of viable infants.

Placenta Praevia. Three cases of central praevia (1.05% incidence) of a large single disc occurred in the nonexperimental pregnancies, two of these being in consecutive pregnancies in one female. Marginal praevia, involving the secondary (anterior) placental disc, was diagnosed in two other pregnancies. Emergency C-section, carried out in one case of central praevia after onset of sudden profuse hemorrhage, resulted in the delivery of a dead fetus. In the remaining two cases of central praevia, and in the marginal praevias, diagnosis was made before hemorrhage occurred (as early as day 116 of gestation in one of the former). The period of greatest risk for placental tearing and hemorrhage is during the initial stages of dilatation and effacement of the internal cervical os. Because of this, females in which the condition was suspected were examined at daily intervals. At the earliest sign of cervical change, C-section was carried out with delivery of live infants.

Initially, placenta praevia was diagnosed by ultrasonic stethoscopy. The absence of placental souffle over the normally expected areas of the uterus (mid-anterior and posterior fundic walls), and its detection over the lower uterine segment were highly suggestive of the condition. Diagnosis was confirmed by cautious examination of the cervix by vaginoscopy. The deep red color of the placental tissue, not to be confused with the sometimes hyperemic cervical colliculi, was readily apparent through the dilated cervical ora.

Uterine Rupture. No case of uterine rupture has occurred in our colony. This rare condition, which requires immediate C-section, is characterized by protracted labor, bright red vaginal hemorrhage, and maternal shock (Lapin and Yakovleva, 1963).

Miscellaneous Conditions. The variety of miscellaneous conditions, marked by uterine bleeding, are of such probable rarity that it would be inappropriate to consider them here.

Prolonged Pregnancy

Before describing the pathological conditions associated with prolonged pregnancy in the rhesus monkey, a warning should be given. In cases of suspected post-term pregnancies, breeding records should first be examined to determine possible errors in designating the time of conception. Even with restricted matings to obtain dated pregnancies, mistakes can be made in determining the cycle of conception. If, after reviewing menstrual records, uncertainty still exists, fetal age can be estimated radiographically from the appearance of ossification centers (van Wagenen and Asling, 1964), by measurement of long bones (Hutchinson, 1966), and by the biparietal diameter of the skull (Ferron, Miller, and McNulty, 1976).

In our colony, the mean length of pregnancies terminating in viable births was 164 ±6.7 days. Less than 10% of pregnancies exceeded 1 S.D. above the mean, and only 2% exceeded 178 days in length. The number of stillbirths included in this study is too small to determine reliably any correlation that might exist between prolonged gestation and causes of fetal death. However, there was a tendency for stillbirths, due to placental insufficiency, to occur beyond 1 S.D. of the mean. Had not surgical delivery been carried out to salvage distressed fetuses, many pregnancies might well have continued beyond 2 S.D. of the mean length. To what extent pregnancy may be prolonged in certain cases of late gestational death of the fetus is not certain.

Protracted or Recurrent Labor

The commonest cause of abnormally protracted or recurring labor was dystocia due to malpresentation or malposition of the fetus. Cephalopelvic disproportion is another potential cause although no proven case of this occurred in our colony. Manual delivery may be impossible if the fetus is tightly wedged in the pelvic canal. In this event, limited embryootomy (if the fetus is dead) or C-section are indicated. Uncorrected, the condition may result in fatal exhaustion of the mother. Presentational and postural abnormalities of the fetus are considered in greater detail below.

One case of protracted labor occurred due to pathological contraction ring. The distressed fetus was successfully delivered by C-section. Constriction of the so-called physiological retraction ring (anatomical internal os) occurs during labor to produce an annular stricture (Bandl's ring) in the lower uterine segment, thereby impeding delivery of the fetus. The contraction ring was readily palpable through the maternal abdominal wall.

Rather than a primary cause of prolonged labor, uterine inertia seemed to occur secondary to fetal dystocia.

Fetal Malpresentation/Malposition/Cephalopelvic Disproportion

Malpresentation, leading to dystocia during labor, was the commonest cause of fetal death in late pregnancy (Table 1). Seven breech deliveries occurred, constituting one-third of the stillbirths in which the cause of death was established. However, malpresentation is potentially dangerous only if parturition is imminent. During the first year of study, when five of the seven stillbirths occurred due to breech dystocias, no attempt was made to correct caudal presentation. Caesarean delivery of two viable infants because of malpresentation was deemed, in retrospect, to have been unwarranted. Later experience indicated that fetal malpresentation could usually be corrected by manual transabdominal version. Three possible exceptions to this are 1) when the caudally presented fetus is fully engaged with the cervix so that manipulation is impossible (prompt surgical delivery would have avoided one stillbirth because of this), 2) repeated reversion of the fetus to the breech presentation after several successive manipulations to correct the presentation, and 3) malpresentation of one or both members of twins (one twin pregnancy was successfully terminated by C-section because of this). Dangers may attend such manipulations, e.g., uterine or placental rupture, premature rupture of the membranes and strangulation of the umbilical cord. However, only one of 45 fetuses died as a result of such manipulation, the cause having been attributed to cord strangulation. Because of this, counting fetal heart rate before and after manual version, to determine whether immediate strangulation of the cord had occurred, became a routine procedure.

Abnormal presentation of the fetus was a feature of placenta praevia in the rhesus monkey. In the cases of central praevia, because of the intervening placenta, the vertex could not be brought into direct contact with the internal cervical os by applying manual pressure over the uterine fundus. In the cases of marginal praevia, the fetus reverted to a breech presentation after repeated manual versions. Presumably, the narrow buttocks could engage more readily than the vertex within the restricted space of the lower uterine cavity.

Malposition in the cephalic presentation caused two fetal deaths during second stage labor. In each case, the occiput was posterior so that the fruntum (forehead) became the presenting part. One of the fetuses was manually delivered. The other was so tightly wedged in the pelvic canal that C-section was required.

No proven case of cephalopelvic disproportion occurred in our colony. Rather, we believe dystocia of a cranially presented fetus is most likely to be due to malposition of the body axis or head.

Umbilical Cord Prolapse/Strangulation

Prolapse of the umbilical cord (an 8 cm loop) through the partially dilated cervix was discovered during routine examination of one female in first stage labor. The first untoward sign noted was a precipitous decrease in fetal heart rate from 180 to 60 beats/min, indicating peracute distress. Altering the mother's position did nothing to improve fetal heart rate. Upon discovering the prolapse, during vaginoscopic examination, attempts were made to replace the umbilical cord. The mother was observed continuously through the course of labor. Although faint heart beats were detected on delivery, the infant died a few minutes later. This case was seriously mismanaged. The correct procedure should have been to relieve the pressure of the fetus' head on the prolapsed cord and then to perform C-section immediately.

One fetus, delivered dead by C-section, was found to have its umbilical cord wrapped around the neck and shoulder. Strangulation probably resulted from repeated manual version of the fetus because of breech presentation (see above).

Symptoms of In Utero "Fetal Distress"

Although listed as a separate entity, in utero fetal distress coexists with many of the signs and conditions enumerated in Table 3. Distress results whenever fetal oxygen supply via the placenta or umbilical cord is severely impaired. Overriding signs, such as profuse uterine hemorrhage or protracted labor, will alert one to the acute or subacute risks imposed on the fetus. However, in the more chronic, as well as in some acute conditions, fetal distress may not be suspected from cursory observation (e.g., as in placental insufficiency and in prolapse or strangulation of the umbilical cord). It is only by close examination that evidence of fetal hypoxia-anoxia will be obtained. Thus, in four pregnancies in which fetal death occurred in utero (Table 1), and in seven in which viable but distressed fetuses were delivered by C-section (Table 2), the only untoward sign was meconium staining of the amniotic fluid (with and without oligohydramnios). Examination of the delivered placentae revealed extensive infarction and necrosis. Marked fetal bradycardia (80 beats/min or less) may also indicate distress, as was found in one case of prolapse of the umbilical cord. One pregnancy was terminated surgically because the high fetal rate, exceeding 240 beats/min, was considered an ominous sign of distress. Later experience indicated, however, that such high rates, while unusual, are not necessarily pathological.

Fetal Death In Utero

The absence of fetal heart beats and blood flow in the fetal placenta, readily confirmed by ultrasonic stethoscopy, is a clear indication of fetal death in utero. In the first instance, attempts should be made to determine the cause of fetal demise. During this study, only one case of late gestational fetal death was attributable to maternal infection, although in earlier stages of pregnancy abortion was a more common consequence (especially due to shigellosis). Maternal anemia (6 mg% hgb), of unknown cause, was associated with one other fetal death in late pregnancy. Metabolic disorders, such as pregnancy (hypertensive) toxemia, have not clearly been shown to occur in rhesus monkeys. However, in two females not considered in this study, traumatic shock received during a severe fight among members of a pen-housed group resulted in death and rapid vaginal expulsion of their fetuses.

Once the cause of fetal death has been established, appropriate treatment of the mother should be immediately instituted. Termination of pregnancy by Caesarean hysterotomy should be considered only if the continued presence of the dead fetus endangers the mother's wellbeing.

Death of fetus before onset of labor may not, in itself, lead to dystocia or endanger the mother's life. However, if the placental membranes have ruptured there is risk of ascending infection of the uterine cavity and placenta. The mother should be monitored daily for hematological evidence of developing infection. Treatment with antibiotics is indicated if infection occurs. Later, when the infection has been controlled, Caesarean delivery may be considered necessary. In the event of resorption or leakage of amniotic fluid, drying of the fetal skin may increase frictional resistance, thereby causing dystocia. Failure to manually correct malpresentation and malposition of a dead fetus may also necessitate Caesarean delivery.

Maternal Hernias and Prolapses

Herniation of the perineum developed in the latter half of pregnancy in two females. Frequent examination revealed that in neither was the fetus compromised and normal deliveries took place. Surgical exploration, 2 weeks after delivery, indicated an attenuation of the sacropelvic muscles and ligaments. The greatest danger to the female was retroversion and entrapment of the bladder with subsequent development of cystitis.

Prolapse of the cervix or vagina occurred in mid-gestation in six females. In four of these, there was evidence of weakened musculature of the perineal region suggestive of a partial perineal hernia. The animals were examined at frequent intervals to determine any deleterious effects on the fetus. In only one was the prolapse considered serious enough to warrant C-section; normal deliveries occurred in the other animals. However, in another pregnancy, not included in this study, death of the fetus in utero probably was a consequence of mechanical interference with normal parturition.

Prolapse or eversion of the endocervical mucosa, through the external os, was not an uncommon finding in rhesus monkeys during pregnancy. It appeared to present no obstetrical problems.

Sudden herniation of the umbilical region, at the site of previous surgical incisions, occurred in two females in late pregnancy. Normal deliveries followed.

DISCUSSION

In the obstetrical management of nonhuman primates, the clinician is sometimes confronted with the difficulty of deciding if and when C-section is necessary. On the one hand, he wishes to avoid unnecessary uterine surgery since this always carries with it the risk of impeding future reproductive function. Yet, on the other hand, failure to act promptly may result in unnecessary death of the mother or fetus. Certain overt signs, such as the profuse uterine hemorrhage associated with placenta praevia or abruptio placenta, indicate the urgency of the situation when there is no choice but to perform emergency hysterotomy. However, cursory, even if frequent, observation of females in late pregnancy may fail to reveal many types of obstetrical complications. Thus, early diagnosis or anticipation of abnormalities, essential to the successful antenatal care of rhesus monkeys, was possible only by periodic examination of the mother and fetus. Among the main objectives of the prepartum examinations were to determine fetal viability, predict parturition, ascertain fetal presentation and evaluate in utero distress of the fetus.

Most important in the monitoring of pregnancy was the introduction of ultrasonic stethoscopy. Prior to its use, viability of the fetus was determined solely on the ability to detect spontaneous or elicited body movements. This proved unreliable and, during the first year of study, resulted in incorrect diagnosis of death and unnecessary Caesarean delivery of three viable fetuses. Similarly, determining the exact time of death of a fetus in relation to parturition or Caesarean delivery was impossible without the use of ultrasonic monitoring. Ultrasonic stethoscopy also proved invaluable in localizing the site of placental attachment, important in the diagnosis of placenta praevia. The technique was also helpful in choosing the site for uterine incision during C-section. Although accurate methods were not available for the quantitative assessment of fetal well-being (e.g., electrocardiography, pO_2, pCO_2 and acid/base determinations) gross abnormalities in fetal heart rate (especially marked bradycardia) were readily detectable by ultrasounding.

Predicting the time of parturition was important since the period of greatest risk to the fetus is during labor (e.g., as in breech dystocia, placental insufficiency, placenta praevia). In the rhesus monkey, the variability in gestation length is too great to permit an accurate prediction of parturition based on calculation alone. Although sometimes misleading, assessing the progressive changes in physical characteristics of the cervix (by rectal palpation and by vaginoscopy) was the most convenient method for determining the imminence of parturition.

Breech dystocia was the commonest known cause of fetal deaths. However, caudal presentation can be regarded as potentially dangerous only if parturition is imminent. Transabdominal manual version of the fetus into the cephalic presentation can be readily achieved unless it has become fully engaged with the cervix. In this event, C-section is advised. Hysterotomy may also be indicated if the fetus repeatedly reverts into the breech following several successive attempts to correct presentation or if one or both members of twins are malpresented. As a precaution, fetal heart rate should be counted before and after manipulation to determine whether strangulation of the umbilical cord may have occurred. Malposition in the cephalic presentation was also a cause of dystocia. Discovery, however, may be too late to avoid death of the fetus.

The other major cause of fetal demise encountered in this study was placental insufficiency. This condition should always be considered in cases of suspected prolonged pregnancy. Meconium staining of the chorioamniotic membranes, however, may be the only adverse sign noted. Caesarean delivery is indicated to avoid the additional stress during labor on an already hypoxic fetus.

Many of the other obstetrical conditions encountered in this study were of infrequent or rare occurrence (e.g., placenta abruptio and placenta praevia, umbilical cord prolapse). Nonetheless, their combined effect on reproductive output over the 2½ years was significant.

The development of maternal perineal hernias and vaginal/ cervical prolapses in late pregnancy may have been an effect of prolonged caging with lack of adequate exercise. Surprisingly, such physical abnormalities appeared not to inhibit normal progress of parturition in most cases. Because of the difficulty, or impossibility, of surgical repair, it might be inappropriate to rebreed such females.

Because of the risks to the fetus, strict attention should be paid to even the slightest clinical signs of maternal infection, anemia, or metabolic disorders. Early diagnosis and immediate treatment are the best approach. Caesarean hysterotomy is indicated only if continuation of pregnancy would seriously jeopardize the mother's health. In the event of prolonged retention of a dead fetus, the mother should be monitored daily for early evidence of developing infection.

A number of fetal deaths occurring in this study could have been avoided by prompt surgical delivery. One case of central placenta praevia had been diagnosed several hours before onset of profuse uterine hemorrhage. Stillbirth occurred only hours after several unsuccessful attempts were made to correct breech presentation of a pelvically engaged viable fetus. Emergency Caesarean delivery during first stage labor probably would have avoided death of another fetus due to prolapse of the umbilical cord. In three pregnant females, meconium staining of the amniotic fluid indicated fetal hypoxia. Rather than allowing labor to ensue, with fatal consequencs to the fetuses, C-section should have been done.

Cases of suspected prolonged or desultory labor are frequently mismanaged in monkeys. Fear of aggravating the situation often leads to the conclusion to "leave well enough alone." One should not hesitate, however, to examine an animal to determine the cause of dystocia. Manual delivery of live fetuses in breech or brow presentations is often difficult. Unless correction of the dystocia can be quickly achieved, it is wiser to resort immediately to C-section. Because of the difficulties of continually monitoring the mother and fetus, induction or promotion of labor by the intramuscular administration of pitocin cannot be recommended. Although successful in some instances, the rapidity of the uterine response increases the risk of uterine spasm, prolapse or compression of the umbilical cord, and development of a compound malpresentation or malposition of the fetus.

Successful Caesarean delivery of fetuses, already suffering varying degrees of in utero hypoxia, depends on minimizing stress to the mother and on a rapid, but careful, surgical procedure. In cases of suspected placental insufficiency, placing the mother in a rich atmosphere of oxygen, while surgical preparations are being made, may or may not improve oxygenation of the fetus. In any event, it will do no harm. The choice of anesthetic is critical. To avoid depressive effects on the fetus, infiltration of the maternal abdominal wall by local blocking agents is a frequently recommended method of anesthesia. However, as Myers (1975) has shown, the stress of handling a conscious mother may have profound deleterious effects on an already distressed fetus. For this reason, general anesthesia was obtained in this study by a divided intramuscular dose of ketamine hydrochloride. In the rhesus monkey, fetal heart rate remains at preanesthetic levels for at least 20 min after administration of this agent (Mahoney and Eisele, 1976), by which time Caesarean delivery of the infant can be accomplished. Upon delivery all attention should be directed towards inducing respiration in the distressed infant. Clamping and severence of the umbilical cord can be delayed without harm, and perhaps with great benefit, to the fetus.

It is probably unrealistic ever to expect perinatal care in laboratory primates to approach the degree of success achieved in human obstetrics, at least in that of the more affluent societies. However, the ever increasing demand for laboratory-bred, rather than wild-caught, nonhuman primates renders the wasteful, and often unnecessary, loss of fetuses scientifically and economically unacceptable. Even with the rather unsophisticated methods used in this study, a substantial decline in fetal mortality was obtained (Table 4).

TABLE 4. INCIDENCE OF VIABLE DELIVERIES, STILLBIRTHS AND ABORTIONS IN A NON-EXPERIMENTAL BREEDING COLONY OF Macaca mulatta DURING CONSECUTIVE 6-MONTH PERIODS

	PRESTUDY PERIOD		STUDY PERIOD				
	I	II	I	II	III	IV	V
Viable Deliveries (%)	82.86	76.92	81.63	86.21	91.89	91.11	89.48
Stillbirths (%)	11.43	17.95	10.20	8.05	8.11	5.56	5.26
Abortions (%)	5.71	5.13	8.16	5.75	0.00	2.22	5.26
Total Number of Pregnancies	35	78	49	87	37	90	38

ACKNOWLEDGMENTS

We gratefully acknowledge Michael Hempell for his dedicated care of the animals, and Denis Mohr for skillful surgical assistance.

REFERENCES

Bernstine, R. L. Placental capacity and its relationship to fetal health. Clin. Obstet. Gynec. 3:852,1960.

Ferron, R. R., Miller, R. S., and McNulty, W. P. Estimation of fetal age and weight from radiographic skull diameters in rhesus monkeys. J. med. Primat. 5:4-48, 1976.

Hendrickx, A. G. Pp. 120-123 in: Teratogenicity Findings in a Baboon Colony: Conference on Nonhuman Primate Toxicology, C. O. Miller (Ed.), Department of Health, Education and Welfare, Food & Drug Administration, 1966.

Hutchinson, T. C. A method for determining expected parturition date of rhesus monkeys (Macaca mulatta). Lab. anim. Care 16:93-95, 1966.

Lapin, B. A. and Yakovleva, L. A. Comparative Pathology in Monkeys, Trans. U.S. Joint Publ. Res. Service, Springfield: Charles C Thomas, 1963.

Mahoney, C. J. Practical aspects of determining early pregnancy, stage of foetal development, and imminent parturition in the monkey (Macaca fascicularis). In: Breeding Simians for Development Biology, F. T. Perkins and P. N. O'Donaghue (Eds.), Lab. Anim. Handb., Vol. 6, 1975.

Mahoney, C. J. and Eisele, S. Use of an ultrasonic blood flow monitor for determining fetal viability in the rhesus monkey (Macaca mulatta). J. med. Primat. 5:284-295, 1976.

Mahoney, C. J. and Eisele, S. A programme of prepartum care for the rhesus monkey Macaca mulatta: Results of the first two years of study. Pp. 265-267 in: Recent Advances in Primatology, Vol. 2: Conservation, D. J. Chivers and W. Lane-Petter (Eds.), New York: Academic, 1978.

Myers, R. E. The pathology of the rhesus monkey placenta. Pp. 221-257 in: Symposium on the Use of Nonhuman Primates for Research on Problems in Human Reproductions, Sukhumi, U.S.S.R., 1971.

Myers, R. E. Maternal psychological stress and fetal asphyxia: A study in the monkey. Amer. J. Obstet. Gynec. 122:47-59, 1975.

Valerio, D. A., Courtney, K. D., Miller, R. L., and Pallotta, A. J. The establishment of a Macaca mulatta breeding colony. Lab. anim. Care 18:589-595, 1968.

van Wagenen, G. Vital statistics from a breeding colony: Reproduction and pregnancy outcome in <u>Macaca</u> <u>mulatta</u>. J. med. Primat. 1:3-28, 1972.

van Wagenen, G. and Asling, C. W. Ossification in the fetal monkey (<u>Macaca</u> <u>mulatta</u>): Estimation of age and progress of gestation by roentgenography. Amer. J. Anat. 114:107-132, 1964.

Vorherr, H. Placental insufficiency in relation to postterm pregnancy and fetal postmaturity: Evaluation of fetoplacental function; management of postterm gravida. Amer. J. Obstet. Gynec. 123:67-103, 1975.

Walker, J. Foetal anoxia. J. Obstet. Gynaec. Brit. Emp. 61:162-180, 1954.

CHAPTER 2

EFFECTS OF PARENTAL RISK AND
PRENATAL STRESS ON PREGNANCY OUTCOME

R. A. Holm

Child Development and Mental Retardation Center
and Regional Primate Research Center
University of Washington
Seattle, Washington

The general hypothesis that anomalies in the environment profoundly affect the conceptus has received extensive support from teratological studies throughout this century (Stockard, 1910; Hale, 1935; Joffe, 1969). Sontag (1941) was one of the first to extend this hypothesis to include variations in emotional status during pregnancy. Thompson (1957) found that administering maternal emotional traumas during pregnancy resulted in increased emotionality in newborn rat pups. His report was followed by a number of other rodent studies, including a wide range of stress and behavioral measures. Most manipulations produced changes in the offsprings' behavior; however, the specificity and directionality of the effects have not been consistent.

Human studies have found that maternal anxiety during pregnancy can be related to obstetric complications (Grimm, 1961; McDonald and Christakos, 1963). In a retrospective study Stott (1957) found a higher incidence of various prenatal stresses--including illness, death in the immediate family and severe marital discord--in a group of mentally retarded offspring than in several control groups of normal offspring. Also, within each group prenatal stress was related to an increase of early childhood illnesses.

In vivo studies have shown dramatic effects of acute psychological stress to the mother on fetal activity in the human (Sontag, 1966) and on fetal heart rate, blood pressure, and blood gases in the monkey (Myers, 1975). We have undertaken a long-term experimental study of the effects of prenatal stress on the gestation and subsequent development of neonatal pigtail macaques (Macaca nemestrina). Only preliminary findings on pregnancy outcomes will be presented here.

METHODS

Subjects were 39 pigtail macaques (Macaca nemestrina) selected as being at particularly high or low risk for bad pregnancy outcome based on their past breeding records (Sackett, Holm, Davis, and Fahrenbruch, 1975). All animals were from the breeding pool of the Primate Field Station of the Regional Primate Research Center at the University of Washington. All females had at least three prior pregnancies while in the colony. The 15 high-risk females had consistent patterns of abortion, stillbirth, neo-natal death, or low birth weight in their offspring, while the 18 low-risk females had no prior history of fetal loss. Three high-risk and three low-risk males were selected, each having at least 20 conceptions and a pregnancy outcome pattern consistent with their respective female risk group.

All breeding was done within risk groups. Each female contributed a pregnancy under the high and low stress conditions with order counterbalanced. Except for breeding, all animals were singly housed for the entire study. Timed matings were achieved by placing breeding pairs together for 30 min on 3 consecutive days when the female's perineum was maximally swollen (Anderson and Erwin, 1976). Pregnancy was determined by using bimanual palpa-tion 28 days after breeding. If the test was negative, the female was rebred on the next cycle. If the test was positive, the female was moved to either the stress or non-stress room (described below) as dictated by experimental design.

From the 28th to 125th days of pregnancy, each female in the stress room was hand-captured 5 days per week, placed in a small transport cage for 3 min, then returned to her cage. The females displayed high levels of disturbance and threat behavior during the handling procedure, and never became accustomed to the proce-dure. In addition, the person administering the capture/restraint procedure entered the room several times a day and randomly captured one or two females. Thus, all animals in the room became disturbed each time he entered. Females in the non-stress room were disturbed and handled much less frequently and only as required for infrequent routine animal care procedures and for occasional experimental blood draws.

Females were housed in the Primate Field Station at Medical Lake for breeding and during most of their pregnancy. At 125 days of pregnancy females were flown to the Infant Primate Research Laboratory in Seattle. All animals were then housed under the low-stress condition and were observed via infrared television at 0.5-hr intervals for signs of labor until delivery occurred.

Blood was drawn from all females for plasma cortisol determinations. Samples were taken before mating and at 28, 56, 84, 112, and 130 days of pregnancy.

RESULTS

To date, 13 low-risk and eight high-risk females have completed the study with a pregnancy under each stress condition. An additional five low-risk and seven high-risk females have completed the first experimental pregnancy, about half under each stress condition. Table 1 displays the outcomes of these 54 pregnancies by risk group and stress. The mixing of paired and independent subjects in this table makes significant testing difficult; however, some trends do stand out. The high-risk group has a higher incidence of abortion than the low-risk group. Stress appears to have no effect on the high-risk females, while dramatically increasing abortion in the low-risk breeding. The latter point is better illustrated in Table 2, which presents the matched outcomes under each stress condition for females who have completed the study. Abortions, stillbirths, and breech-presentation deaths have all been pooled as fetal loss. None of the high-risk females changed pregnancy outcome from one condition to the

TABLE 1.   OUTCOME OF 54 PREGNANCIES BY RISK GROUP
            AND STRESS

	HIGH RISK		LOW RISK	
	Stress	No Stress	Stress	No Stress
Abortion	42% (5)	55% (6)	33% (5)	6% (1)
Breech-Stillborn	8% (1)	9% (1)	7% (1)	0% (0)
Liveborn	50% (6)	36% (4)	60% (9)	94% (15)
N =	(12)	(11)	(15)	(16)

next. The five low-risk females whose outcomes were different under the two conditions all had liveborns in the low-stress condition and fetal losses under the high stress. A one-tailed binomial test suggests that this is a reliable effect (for 0 out of 5, p = 0.031).

The plasma cortisol determinations made to date suggest that no differences in plasma cortisol between risk or stress groups or between pregnancy-outcome groups will be detected by this technique.

The gestational ages at which abortions occurred are presented in Table 3. Almost all the abortions occurred during the first half of pregnancy. The four groups displayed no remarkable differences in gestational age of the abortions. The repeat

TABLE 2. MATCHED OUTCOMES FOR EACH STRESS CONDITION

HIGH RISK	Fetal Loss	Live	LOW RISK	Fetal Loss	Live
No Stress	5	3	No Stress	0	13
Stress	5	3	Stress	5	8

TABLE 3. DAY OF GESTATION ON WHICH ABORTION OCCURRED

	LOW RISK			HIGH RISK	
Subject	Stress	No Stress	Subject	Stress	No Stress
1	57	--	1	54	58
2	44	--	2	109	--
3	116	--	3	--	37
4	42	--	4	89	75
5	85	--	5	44	48
6	--	79	6	65	--
			7	76	77

aborters were very consistent as to when they aborted their fetuses (e.g., subjects 1, 4, 5, and 7 from the high-risk group).

DISCUSSION

The failure of the cortisol measures to pick up differences between the stressed and non-stressed females is surprising, since the animals showed marked behavioral responses to frequent handling. Furthermore, the increased abortion rate seen in the low-risk, stressed group indicates that the procedures are effective. We suspect that the plasma cortisol values we were taking are not a good indicator of chronic stress. More stable measures might be 12- or 24-hr urine-free cortisol or maximum cortisol output in response to ACTH challenge which could indicate adrenal hypertrophy in the stressed animals. We have begun taking 12-hr urine samples at 112 days post-conception and plan to initiate the ACTH challenge shortly.

A majority of abortions occur at the time the placenta is taking over hormonal control of the pregnancy from the ovaries. We are preparing for additional studies concentrating on the endocrine changes occurring during this period, and hope to find some of the specific mechanisms causing the abortions. The consistent timing of repeat aborters affords us an excellent opportunity to do detailed endocrine studies during the period of expected abortion.

The most notable finding of this study to date is the interaction of the risk and stress conditions. Many of the high-risk females abort independently of our stress manipulations. The low-risk females, if they abort at all, do so almost exclusively under the stressed condition. Increased abortions would be predicted from Myers' (1975) findings that in restrained rhesus monkeys with indwelling maternal and fetal arterial catheters, the mere entry of the experimenter after an hour or so of absence resulted in marked changes in fetal heart rate, blood pressure, and blood gas. Heightened fetal asphyxia during these episodes of maternal psychological stress was often severe enough to compromise the fetus. The dramatic increase in abortion rate in the low-risk, stressed group does not, however, coincide with the clinical impressions of many veterinarians who detect no increase from routine manipulations of pregnant monkeys for clinical or research needs.

We believe most of these clinical manipulations are less frequent than ours and are concentrated later in pregnancy than when most of our abortions occur.

ACKNOWLEDGMENTS

This research was supported by NIH grants HD08633 and HD02274 from NICHD Mental Retardation Branch and RR00166 from the Animal Resources Branch.

REFERENCES

Anderson, B. and Erwin, J. A comparison of two timed-mating strategies for breeding pigtail monkeys (Macaca nemestrina). Theriogenology 4:153-156, 1976.

Grimm, E. R. Psychological tension in pregnancy. Psychosom. Med. 23:520-527, 1961.

Hale, F. Relation of vitamin A to anophthalmos in pigs. Amer. J. Ophthal. 18:1087-1093, 1935.

Joffe, J. M. Prenatal determinants of behavior. International Series of Monographs in Experimental Psychology, London: Pergamon Press Ltd., 1969.

McDonald, R. L. and Christakos, A. C. Relationship of emotional adjustment during pregnancy to obstetric complications. Amer. J. Obstet. Gynec. 86:341-348, 1963.

Myers, R. E. Production of fetal asphyxia by maternal psychological stress. Pp. 3-15 in: Progress in EPH-Gentosis, E. T. Rippman, H. Stamm, H. P. McEwan, and P. Howie (Eds.), Basel: Organization Gestosis-Press, 1975.

Sackett, G. P., Holm, R. A., Davis, A. E., and Fahrenbruch, C. E. Prematurity and low birth weight in pigtail macaques: Incidence, prediction, and effects on infant development. Pp. 189-205 in: Proceedings of the 5th Congress of the International Primatological Society, Tokyo: Japan Science Press, 1975.

Sontag, L. W. Significance of fetal environmental differences. Amer. J. Obstet. Gynec. 42:996-1003, 1941.

Sontag, L. W. Implications of fetal behavior and environment for adult personalities. Ann. N.Y. Acad. Sci. 134:782-786, 1966.

Stockard, C. R. The influence of alcohol and other anesthetics on embryonic development. Amer. J. Anat., 10:369-392, 1910.

Stott, D. H. Physical and mental handicaps following a disturbed pregnancy. Lancet, May 18:1006-1012, 1957.

Thompson, W. R. Influence of prenatal maternal anxiety on emotionality in young rats. Science 125:698-699, 1957.

CHAPTER 3

GROSS PLACENTAL MORPHOLOGY AND PREGNANCY OUTCOME IN Macaca nemestrina

C. E. Fahrenbruch, T. M. Burbacher, and
G. P. Sackett

Infant Primate Research Laboratory of the
Child Development and Mental Retardation Center
and Regional Primate Research Center
University of Washington
Seattle, Washington

INTRODUCTION

Various measures have been suggested to estimate placental functional efficiency. Gruenwald and Minh (1961) used placental weight as one index of placental function in humans. They found that variation in placental weight is not significant and concluded that fetal size is not determined by the placenta in normal pregnancy. Placental weight is a poor predictor of birth weight (Gruenwald and Mihn, 1961; Thomson, Billewicz, and Hytten, 1969). Thomson et al. (1969), reviewing the relation between placental weight and birth weight in normal and abnormal pregnancies, concluded that placental insufficiency due to small size is rare. Fetal-placental weight ratios may correlate with placental efficiency. Lubchenco (1976) indicated that this measure is based on the assumption that the growth-retarded fetus is limited by small placental size and may have a greater fetal-placental weight ratio. Another functional measure, probably correlated with placental weight, is trophoblastic surface or mean villous area (Aherne and Dunnill, 1966). During gestation the trophoblastic surface increases and the membrane becomes thinner, suggesting greater placental efficiency near term. In a study of the relation between placental insufficiency and the small-for-date baby, Scott and Jordan (1972) incorporated general placental appearance, cord and membrane characteristics, abnormalities, histologic assessment,

and weight into a scoring system. The weighted scores discrimi-
nated somewhat between cases of placental insufficiency and
controls. The present study describes the gross morphology of 59
pigtail macaque (Macaca nemestrina) placentas and, in a smaller
sample of term-gestation placentas, examines the relation between
placental characteristics and pregnancy outcome.

METHODS

Fifty-nine M. nemestrina placentas were obtained over a 2-
year period from 14 Caesarean section and 45 vaginal deliveries in
our laboratory. Thirty-six placentas were from timed matings,
with a gestation range of 115-187 days. For vaginal deliveries the
mothers were anesthetized with ketamine hydrochloride administered at a dosage of 10 mg/kg of body weight immediately after
giving birth to facilitate the recovery of the placenta and the
examination of the infant. The gross placental morphology was
recorded as soon as possible after delivery. The umbilical cord and
fetal membranes were removed and lobe weights and dimensions
were recorded.

RESULTS

Gross Placental Morphology

Eight of the 59 placentas were single-lobed. Four of the
eight single-lobed placentas had either circummarginate or circum-
vallate membrane insertions, whereas 12 of the 51 (25%) double-
lobed placentas had abnormal membrane insertions. All 59 umbilical cords had two umbilical arteries and one umbilical vein. One
umbilical cord was knotted. The primary lobe, where the umbilical
cord inserts, was usually the larger, although the ratio of the two
weights varied widely. Four of the 51 double-lobed placentas had
heavier secondary lobes. Two placentas showed partial lobe fusion.
Recent retroplacental and retromembranous blood clots and areas
of infarction were common. Two placentas showed evidence of
pre-term separation from the uterine wall as indicated by old
retroplacental clots. No instances of placenta praevia or meconium staining were noted. No difference was found between the
weights of single- and double-lobed placentas, controlled for gestation. However, the variability was great, particularly in the small
pre-term Caesarean-section delivery sample. Figure 1 illustrates
the distribution of placentas by gestation and placental weight.
The mean weight of the short gestation placentas was 90 g, with a

range of 64-121 g. The mean weight of the term placentas was 125 g, with a range of 91-179 g. Time of placenta delivery varied from immediately after birth to 8 hr after birth. Most were delivered within an hour after birth.

Placental Characteristics and Pregnancy Outcome

Table 1 compares the placental and delivery characteristics of five stillborn infants with those of 31 liveborn infants. These 36 infants were of known gestational age and were compared on the

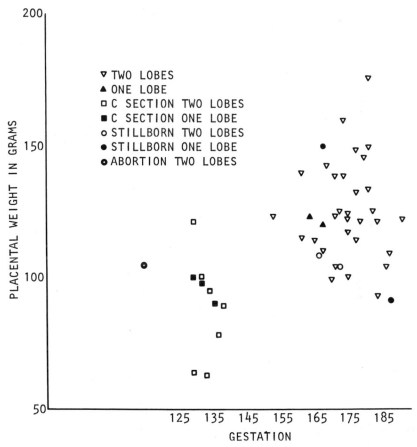

Figure 1.   Distribution of placentas by gestation and placental weight.

TABLE 1.   PLACENTAL CHARACTERISTICS BY PREGNANCY
            OUTCOME IN INFANTS OF TERM GESTATION

Characteristic	Liveborn Total N=31	Stillborn Total N=5
single lobe	2	2
non-central umbilical cord insertion	7	4
infarcts	26	3
retroplacental hemorrhage	29	4
retromembranous hemorrhage	13	1
breech presentation	3	4

basis of normal gestation, from 162 to 182 days. No relation was found between pregnancy outcome and the variables placental weight, lobe number, membrane insertion, and birth weight. The relation between breech presentation, noncentral umbilical cord insertion, and pregnancy outcome was striking, although unclear. The relative odds of stillbirth were 37 times greater for infants delivered in a breech presentation. An even higher risk estimate is suggested from other data collected at the Infant Primate Research Laboratory as delivery teams assisted in the delivery and resuscitation of several breech infants. The relative odds of stillbirth were 13 times greater for infants with noncentral umbilical cord insertion. However, three of the stillborn infants with noncentral cord insertion were also delivered in the breech presentation. The association between noncentral umbilical cord insertion and pregnancy outcome may be manifested through intervening variables such as delivery presentation.

A relation between placental abnormalities and consequent decreased placental efficiency might be examined through other measures of pregnancy outcome. Table 2 compares the birth weights of normal gestation males by umbilical cord insertion. (Only males were compared due to the sex differences in birth weight and smaller number of females with noncentral cord insertion in our sample.) Although the mean birth weight of males with a noncentral umbilical cord insertion was somewhat lower, the difference was not significant.

TABLE 2.   RELATION BETWEEN UMBILICAL CORD INSER-
TION AND BIRTH WEIGHT IN FULL-TERM MALES

Type of Insertion	No. Animals	Mean Birth Weight (g)	S.E.
LIVEBORN MALES			
Central	10	596.5	21.08
Eccentric	5	538.0	13.84
Marginal	1	610.0	---
ALL MALES			
Central*	11	597.3	19.08
Non-central**	7	532.9	22.09

*includes 1 stillborn, breech presentation
**includes 1 stillborn, normal presentation

In the term-gestation group total placental weight, lobe number, and membrane insertion did not significantly alter birth weight when the covariates gestation and sex were first considered. The sex-specific mean values of total placental weight, lobe weight, and dimensions are shown in Table 3. Weight and dimensions of the primary lobe were very similar for males and females. However, secondary lobe weight, length and width were significantly greater for males than for females (Figure 2). The overall placenta weight was also greater for males. Because of the small number and variability in gestation length, the short gestation placenta data were not analyzed for placental characteristics by sex and birth weight.

Very large, old blood clots indicating separation of the placenta from the uterine wall during the course of pregnancy were found in two placentas. One infant was stillborn and of normal birth weight and gestation, suggesting that its breech presentation was the cause of death. The other infant was delivered by a malnourished mother after a prolonged gestation. The infant was small-for-date with a wasted appearance as described by Clifford (1957) in human infants with failing placental function. The placenta was thin and flat with poorly defined cotyledons, a large number of infarctions, and a marginal umbilical cord insertion. These findings are characteristic of fetal growth retardation and accompanying placental insufficiency, probably related to the illness of the mother.

TABLE 3.   CHARACTERISTICS OF DOUBLE LOBED PLACEN-
TAS BY SEX FOR TERM-GESTATION M. nemestrina
INFANTS

	Females (N=14)		Males (N=19)	
	Mean	S.E.	Mean	S.E.
Total weight*(g)	118.08	3.39	129.80	4.67
Wt. of primary lobe (g)	74.58	4.44	77.37	3.61
Length of primary lobe (mm)	110.01	3.18	115.63	3.11
Width of primary lobe (mm)	85.25	3.58	92.26	3.32
Wt. of secondary lobe*(g)	4.358	3.72	52.42	2.68
Length of secondary lobe*(mm)	85.50	3.62	95.37	2.31
Width of secondary lobe**(mm)	65.58	3.02	79.42	2.96

*significant at the 0.05 level by multiple regression
**significant at the 0.01 level by multiple regression

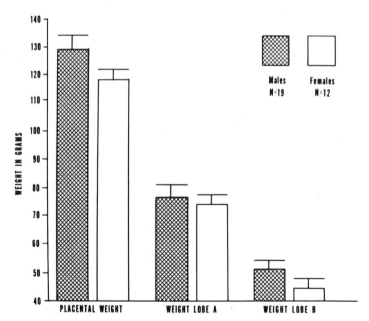

Figure 2.   Double lobe placental weight for liveborn M. nemestrina infants of normal gestation.

Placental Weight and Maternal Characteristics

Although Hendricks (1964) reported human placenta weight to be less for primiparous mothers, the four primiparas in our sample had placenta weights well within the range exhibited by 33 multiparous females. No relation was found between maternal prepregnant weight and placenta weight after controlling for the effects of length of gestation and sex of the infant.

DISCUSSION

The effects of placental abnormalities on fetal growth and pregnancy outcome in the rhesus monkey have been reported by Myers, Hill, Holt, Scott, Mellits, and Cheek (1971). They found an increased incidence of placental abnormalities in single-lobed compared with double-lobed placentas. Umbilical cord and membrane insertion abnormalities were most common in single-lobed placentas. Myers (1972) stated that abnormal insertions may affect pregnancy outcome by decreasing placental function.

In our sample no relation was found between the number of lobes or membrane insertion and pregnancy outcome. However, the number of single-lobed placentas was small, and these findings may not be contradictory. The relation between umbilical insertion and pregnancy outcome may be due to an intervening variable leading to delivery in a breech position. Scott and Jordan (1972) reported an association between placental insufficiency and a thin umbilical cord. Gruenwald (1963) found a relation between cord wrinkling and insufficiency. The association between umbilical cord insertion, delivery presentation, and pregnancy outcome needs to be explored further.

In our sample, placental weight did not correlate with birth weight after adjusting for gestation and sex. This finding may be due to small sample size since only known-aged, term-gestation infants were included in the analysis. In other studies, placental weight has been associated with birth weight in both humans and rhesus monkeys (Thomson et al., 1969; Myers et al., 1971). In humans, placental weight is also related to maternal weight and weight for height and parity, although the association is slight (Hendricks, 1964; Thomson et al., 1969).

ACKNOWLEDGMENTS

We wish to thank members of the delivery teams who examined the placentas, often in the middle of the night. In addition, we thank Dr. Ronald DiGiacomo who assisted us with the placenta examination procedures and checksheet. This work was supported by NIH grant RR00166 to the Regional Primate Research Center, NICHD grants HD08633 and HD02274 to the Child Development and Mental Retardation Center, and SCOR grant HL19187 to the Department of Pediatrics at the University of Washington.

REFERENCES

Aherne, W. and Dunnill, M. S. Quantitative aspects of placental structure. J. Path. Bact. 91:123, 1966.

Clifford, S. H. Postmaturity. Pp. 13-63 in: Advances in Pediatrics, Chicago: Year Book Publications, Inc., 1957.

Gruenwald, P. Chronic fetal distress and placental insufficiency. Biol. Neonat. 5:215, 1963.

Gruenwald, P. and Mihn, H. N. Evaluation of body and organ weights in perineal pathology. II. Weight of body and placenta of surviving and of autopsied infants. Amer. J. Obstet. Gynec. 82:312-319, 1961.

Hendricks, C. H. Patterns of fetal and placental growth: The second half of normal pregnancy. Obstet. Gynec. 24:357-365, 1964.

Lubchenco, L. O. The high risk infant. Pp. 193-195 in: Major Problems in Clinical Pediatrics, Vol. XIV, Philadelphia: W. B. Saunders Co., 1976.

Myers, R. E. The gross pathology of the rhesus placenta. J. Reprod. Med. 9:171-198, 1972.

Myers, R. E., Hill, D. E., Holt, A. B., Scott, R. E., Mellits, E. D., and Cheek, D. B. Fetal growth retardation produced by experimental placental insufficiency in the rhesus monkey. Biol. Neonat. 18:379-394, 1971.

Scott, J. M. and Jordan, J. M. Placental insufficiency and the small for dates baby. Amer. J. Obstet. Gynec. 113:823-832, 1972.

Thomson, A. M., Billewicz, W. Z., and Hytten, F. E. The weight of the placenta in relation to birthweight. J. Obstet. Gynaec. Brit. Cwlth 76:865, 1969.

CHAPTER 4

SEROLOGICAL MATERNO-FETAL INCOMPATIBILITY
IN NONHUMAN PRIMATES

W. W. Socha and J. Moor-Jankowski

Primate Blood Group Reference Laboratory and
World Health Organization Collaborating Centre for
  Haematology of Primate Animals
Laboratory for Experimental Medicine and
  Surgery in Primates (LEMSIP)
New York University School of Medicine
New York, New York

The pathogenesis of erythroblastosis fetalis in man seems to be well known and understood (Wiener and Wexler, 1950; Wiener, 1961). In the classic case, an Rh-negative pregnant woman becomes sensitized to the Rh factor, usually as a result of transplacental leakage of fetal Rh-positive blood into her circulation. The resulting maternal Rh antibodies readily pass through the placenta into the fetal circulation and then coat the red cells of the fetus, leading to their destruction and giving rise to manifestations of hemolytic disease. In the most severe cases, there is stillbirth or hydropic or macerated fetus with marked hepatosplenomegaly; less severely affected offspring are liveborn with pronounced jaundice (icterus gravis) which may lead to kernicterus, causing severe mental retardation. The mildest manifestation is hemolytic anemia. As can be expected, the severity of the manifestations is usually correlated with the degree of sensitization of the mother, i.e., the higher the titer of the maternal antibodies, the greater the stillbirth rate.

Erythroblastosis fetalis occurs naturally in various animal species (Roberts, 1957). In horses, pigs and dogs, the maternal isoantibodies cannot pass through the placental barrier, but appear in the colostrum and thus affect the incompatible offspring only after birth, when first nursing. A situation more similar to that in man occurs in rabbits in which maternal isoantibodies pass through the placenta and produce erythroblastosis in utero (Kellner and Hedal, 1952, 1953). In nonhuman primates, the possibility of erythroblastosis fetalis had been predicted on the basis of their fetoplacental similarity to man as well as the known polymorphism of simian red cell antigens whose immunogenicity has been proved experimentally on many occasions (Moor-Jankowski and Wiener, 1972).

The first nonhuman primate case of erythroblastosis fetalis was described by Gengozian, Lushbaugh, Humason, and Kniseley (1966) in marmosets (Tamarinus nigricollis). Relatively high frequency of materno-fetal incompatibility in these habitually twinning and chimeric primate species made the study of erythroblastosis in marmosets particularly interesting and fruitful. It is noteworthy that the incidence of materno-fetal serologic conflict seems to be higher in offspring of matings between marmoset species, which is comparable to the occurrence of hemolytic disease in mule foals.

Materno-fetal incompatibility has been observed in a large breeding colony of hamadryas baboons (Papio hamadryas) maintained at the Sukhumi Primate Center, USSR (Volkova, Vyazov, Verbitsky, Lapin, Kuksova, and Andreyev, 1966; Verbickij, 1972). Retrospective study of 1,127 pregnancies in Sukhumi baboons showed that incompatible matings yielded "pathological outcome" in almost 16% of the cases, compared with only slightly over 7% in compatible matings. The frequencies of abortions and stillbirths were also twice as high in pregnancies from incompatible matings as from compatible matings. Mortality up to one year of age was significantly higher among offspring of incompatible matings. In 19 cases, spontaneous hemolytic disease of the newborn hamadryas baboons could be directly referred to the sensitization of the mother with the fetal erythrocyte antigens.

The role of materno-fetal incompatibility in the outcome of pregnancy suggested by statistical data and clinical observation has been demonstrated by experimental isoimmunization of female baboons with the red cells of their incompatible breeding mates (Verbitsky, Volkova, Kuksova, Lapin, Andreyev, and Gvazava, 1969). In that study of 43 blood group incompatible full-term offspring of experimentally isoimmunized mothers, there were 17 stillbirths and 26 live-born babies, of which 11 had severe icterus and 8 had anemia. Anemia, hyperbilirubinemia (up to 30 mg%), erythroblastosis of the peripheral blood, erythroid hyperplasia of the bone marrow, extramedullary hematopoiesis in the liver, spleen and lymph nodes, and positive direct antiglobulin test in the affected infants confirmed the identity between this syndrome in hamadryas baboons and erythroblastosis fetalis in man. It is interesting to note that as many as seven offspring were clinically unaffected.

In our unpublished studies, however, we were unable to confirm the above results. Four pregnant olive (P. cynocephalus) and hamadryas baboon females (previously isoimmunized and known to produce high-titer antibodies) were hyperimmunized in the first half of pregnancy with the red cells of their breeding males. All pregnancies resulted in delivery at term of normal infants without clinical symptoms of erythroblastosis fetalis, although in two cases the infants' red cells were moderately coated with maternal antibodies. Similar experiments were carried out with four hyper-immunized female rhesus macaques (Macaca mulatta); again, the hemolytic disease failed to develop even though fetal red cells were maximally antibody-coated. Sullivan et al. (1972) also reported that rhesus infants whose red cells had been maximally coated with antibodies did not show any clinical signs of hemolytic disease.

In another series of experiments, we tried to produce erythroblastosis fetalis in crabeating macaques (M. fascicularis) (Wiener, Socha, Niemann, and Moor-Jankowski, 1975). The trans-placental passage of high-titer maternal antibodies could be estab-lished in this species, as well as the maximal coating of the fetal red cells (as judged from strongly positive direct antiglobulin test and the presence of excess free antibodies in the infant's serum); however, the fetal coated red cells remained intact and no clinical symptoms of intravascular hemolysis were observed.

These and observations noted above indicate that the macaques, as well as some of the baboons that failed to develop clinical symptoms of hemolytic disease despite the presence of maternal antibodies on the offsprings' red cell surface, must benefit from some kind of protective mechanism. The nature of this mechanism is unknown, although some explanations have been proposed to account for similar but extremely rare cases described in man. It could be a polymorphic hereditary inhibitor or an occasional physiologic state, as, for instance, the adequate hydra-tion of the newborn baby which prevents clumping of coated red cells as shown by in vitro tests (Wiener and Wexler, 1963).

In contrast to baboons and macaques, two newborn chimpan-zees showed clinical symptoms of erythroblastosis fetalis; their early deaths could be ascribed to transplacental passage of mater-nal antibodies directed against antigens on their red cells (Wiener, Socha, and Moor-Jankowski, 1977). In a third unpublished case, a previously immunized chimpanzee gave birth to a baby with severe jaundice and high bilirubin level. The red cells of both infant and father were strongly agglutinated by the mother's serum. Free maternal antibodies were also found in the baby's circulation. The maternal agglutinins were found to detect the so-called G^C specificity on the father's and baby's red cells. Significantly, the red cells of the same specificity were used a few years ago for immunization of the mother.

Aside from perinatal observations, the importance of the materno-fetal incompatibility is ascertained from relatively frequent occurrence of so-called natural antibodies (i.e., antibodies not resulting from deliberate immunization) in the sera of randomly screened primate animals. According to Volkova et al. (1966), isoantibodies frequently found in the sera of multiparous macaques resulted from naturally incompatible pregnancies. Our observations seem to confirm that view. Screening the sera of 340 rhesus monkeys maintained for the National Institutes of Health, Division of Research Resources, revealed spontaneous antibodies in about 15% of the samples tested; the great majority of antibody-containing sera were from females (unpublished). In a series of 52 bonnet macaques (M. radiata) maintained at the California Primate Research Center at Davis, four contained natural antibodies on their sera; of these, two showing the strongest activity were obtained from multiparous females (Socha, Moor-Jankowski, Wiener, Risser, and Plonski, 1972). Transplacental immunization was most probably the source of strong natural agglutinins found in the sera of several baboons, as well as a possible source of antibodies in the sera of four among 18 nonimmunized crabeating macaques that we have tested (Socha, Wiener, Moor-Jankowski, Scheffrahn, and Wolfson, 1976).

We have found spontaneously occurring agglutinins in various Old World monkey species which may act in various ranges of temperatures as isoagglutinins, heteroagglutinins (both species-specific and type-specific) and even as autoagglutinins. For all practical purposes, the most important are isoantibodies reactive at body temperature which, as IgG immunoglobulins, readily cross the placental barrier and, therefore, most often result from materno-fetal incompatibility. Although the strength and avidity of such antibodies are generally only moderate (usually the titers reach 8-16 units by antiglobulin method), even these relatively weak agglutinins may cause untoward reactions when transfusion of incompatible blood is given to an animal whose serum contains agglutinating antibodies.

In breeding, the existence of spontaneous agglutinins should be taken into consideration when selecting prospective breeding

TABLE 1.   BLOOD TYPING TESTS THAT CAN BE PERFORMED AT THE PRIMATE BLOOD GROUP REFERENCE LABORATORY.

Species	Human-Type		Simian-Type	
	Systems	Blood Factors	Systems	Blood Factors
Chimpanzee (Pan troglodytes & Pan paniscus)	A-B-O M-N Rh-Hr	A,A_1,H Secretor type Rh_o,hr'	V-A-B C-E-F L-P	V^C,A^C,B^C,D^C X^C,Y^C C^C,c^C,E^C,F^C L^C,P^C G^C,H^C,K^C,N^C,O^C
Gibbon (Hylobates lar)	A-B-O A-B-O M-N Rh-Hr	A,A_1,B,H Secretor Type hr'		A^g,B^g,C^g and 3 unnamed factors
Orangutan (Pongo pygmaeus)	A-B-O	A,A_1,B,H Secretor type		
Gorilla (Gorilla gorilla gorilla & G.g. beringei)	A-B-O M-N Rh-Hr	B Secretor type Rh_o,hr'	Cross reacts selectively with chimpanzee C-E-F reagents	
Baboon (Papio sp)	A-B-O	A-B-H Secretor		$A^P,B^P,C^P,D^P,$ $G^P,H^P,N^P,Q^P,$ Y^P,Z^P,ca,hu & 11 unnamed factors
Gelada (Theropithecus gelada)	A-B-O	Secretor type		3 unnamed factors
Rhesus, crabeating & pigtailed macaque (Macaca mulatta, M. fascicularis and M.nemestrina)*	A-B-O	Secretor type	D^{rh}	$D^{rh},D^{rh}1,D^{rh}2,$ $A^{rh},B^{rh},C^{rh},E^{rh},$ $F^{rh},G^{rh},H^{rh},I^j,$ $X^{rh},U^{rh},Z^{rh},R^j,J^{rh}$ L^{rh},M^{rh},N^{rh} & 5 unnamed factors
Other Old World Monkeys	Human-types			

*Antisera detecting all 24 factors were produced by immunization of rhesus monkeys and show polymorphism in rhesus, crabeating, and pigtailed macaques.

mates. Based on the analogy to human erythroblastosis fetalis, one must consider the possibility of the boosting effect of incompatible pregnancy on the titer of pre-existing antibodies with devastating results to the fetus. It is therefore recommended that blood grouping or at least cross matching tests be performed between prospective breeders to eliminate incompatible matings, especially when dealing with multiparous females.

Serological screening of pregnant females seems indispensable in cases of intrauterine experiments involving supportive blood transfusions. As shown earlier in this chapter, pregnant females are to be considered high-risk blood recipients because of relatively frequent occurrences of natural agglutinins in their sera.

When agglutinins are discovered in the serum of a pregnant female capable of agglutinating the red cells of the breeding mate, periodic titrations of her serum, particularly in the second half of pregnancy, are recommended. The rise of the titer of IgG agglutinins may be indicative of materno-fetal conflict and, in such cases, life-saving exchange transfusion can be envisaged after the infant is delivered with clinical and laboratory symptons of severe erythroblastosis fetalis. In our own practice we were able to save an erythroblastotic newborn chimpanzee by exchange transfusion with thoroughly washed and resuspended mother's red cells (unpublished).

Although not all primate species have been studied for their blood groups, sufficient data have been obtained on serology of species commonly used for research and breeding. Reliable reagents are also available both for human-type and simian-type blood grouping work in these species. On the other hand, screening the sera for the presence of natural antibodies is possible even in the absence of detailed knowledge of the blood group serology of the given primate species. The screening techniques are well established and necessary reagents are available (Sullivan et al., 1972; Erskine and Socha, 1978). The Primate Blood Group Reference Laboratory at LEMSIP is adequately equipped to perform such tests and is willing to carry them out when needed.

Table 1 shows the scope of blood group systems and blood factors in various simian species for which the tests can be carried out at the laboratory.

ACKNOWLEDGMENTS

This research was supported in part by U.S. Public Health Service grant GM12074 and contract RR-4-2184.

REFERENCES

Erskin, A. G. and Socha, W. W. The Principles and Practice of Blood Grouping, 2nd Edition, St. Louis: C. V. Mosby, 1978.

Gengozian, N., Lushbaugh, C. C., Humason, G. I., and Kniseley, R. M. Erythroblastosis fetalis in the primate Tamarinus nigricollis. Nature, Lond. 209:731-732, 1966.

Kellner, A. and Hedal, E. F. Experimental erythroblastosis fetalis. Amer. J. Path. 28:539-542, 1952.

Kellner, A. and Hedal, E. F. Experimental erythroblastosis fetalis in rabbits. I. Characterization of a pair of allelic blood group factors and their specific immune isoantibodies. J. exp. Med. 97:33-49, 1953.

Moor-Jankowski, J. and Wiener, A. S. Red cell antigens in primates. Pp. 270-371 in: Pathology of Simian Primates, R. N. Fiennes (Ed.), Basel: S. Karger, 1972.

Roberts, G. F. Comparative Aspects of Haemolytical Disease, London: Heinemann, 1957.

Socha, W. W., Moor-Jankowski, J., Wiener, A. S., Risser, D. R., and Plonski, H. Blood groups of bonnet macaques (Macaca radiata) with a brief introduction to seroprimatology. Amer. J. phys. Anthrop. 45:485-492, 1972.

Socha, W. W., Wiener, A. S., Moor-Jankowski, J., Scheffrahn, W., and Wolfson, S. K., Jr. Spontaneously occurring agglutinins in primate sera. Int. Arch. Allergy Appl. Immunol. 51:656-670, 1976.

Sullivan, P., Duggleby, C., Blystad, C., and Stone, W. H. Transplacental isoimmunization in rhesus monkey. Fed. Proc. 31:792, 1972.

Verbickij, M. S. The use of hamadryas baboons for the study of the immunological aspects of human reproduction. Pp. 492-505 in: The Use of Nonhuman Primates in Research on Human Production, E. Diczfalusy and C. C. Standley (Eds.), Stockholm: WHO Research and Training Centre on Human Reproduction, Karolinska Institutet, 1972.

Verbitsky, M., Volkova, L., Kuksova, M., Lapin, B., Andreyev, A., and Gvazava, I. A study of haemolytic disease of the foetus and the newborn occurring in Hamadryas baboons under natural consitions. Z. Versuchstierk II:136-145, 1969.

Volkova, L. S., Vyazov, O. E., Verbitsky, M. S., Lapin, B. A., Kuksova, M. I., and Andreyev, A. V. Experimental reproduction of the hemolytic disease of the newborn in Papio hamadryas (linneus). Rev. roum. Embryol. Cytol. 3:119-130, 1966.

Wiener, A. S. Hr-Hr Blood Types, New York: Grune and Stratton, 1961.

Wiener, A. S., Socha, W. W., and Moor-Jankowski, J. Erythro-
    blastosis models. II. Materno-fetal incompatibility in chim-
    panzee. Folia primat. 27:68-74, 1977.
Wiener, A. S., Socha, W. W., Niemann, W., and Moor-Jankowski, J.
    Erythroblastosis models. A review and new experimental
    data in monkeys. J. med. Primatol. 4:179-187, 1975.
Wiener, A. S. and Wexler, I. B. Erythroblastosis foetalis und
    Blutaustausch. Stuttgart: Georg Thieme, 1950.
Wiener, A. S. and Wexler, I. B. Rh-Hr Syllabus, 2nd Ed., p. 51,
    New York: Grune and Stratton, 1963.

CHAPTER 5

PRENATAL PROTEIN AND ZINC MALNUTRITION IN THE RHESUS MONKEY, Macaca mulatta

D. Strobel[*], H. Sandstead[+], L. Zimmermann[*], and A. Reuter[*]

[*]University of Montana
Missoula, Montana

[+]USDA Human Nutrition Laboratory
Grand Forks, North Dakota

INTRODUCTION

Many nutrition researchers have increasingly followed the supposition that malnutrition may cause morphological and bio-chemical damage to the brain (for reviews see Birch and Gussow, 1970; Cravioto and DeLicardie, 1973; Latham, 1974; Read, 1973). This concern has centered on a period of vulnerability in human brain development roughly 3 months before and 6 months after birth, which is generally thought to be the period of maximal brain growth.

There is considerable evidence from both animal and human studies to support the contention that malnutrition in early life can produce irreversible structural changes in the nervous system. Relating these structural changes to alterations in behavioral functions, however, has been less clear.

The major problem in attempting to assess the relation between malnutrition and behavior is the myriad of other variables that may interact to produce such changes. These variables include a variety of differences in genetic and environmental factors. The relative contribution of each of these variables has never been fully articulated in humans because they are hopelessly intertwined, and for ethical and practical reasons have never been successfully separated experimentally.

Animal models of malnutrition have been useful in eluci-
dating the impact of early malnutrition. However, there is reason
to believe that there are limits to the application of these models
to the human condition. For example, the exact periods of
vulnerability may be quite different among species of animals.
More importantly, there may be major differences in the adaptive
mechanisms available to the organism to withstand a nutritional
insult.

In addition, studies of animals or humans must differentiate
between damage to learning capacities and alterations in learning
performance. In other words, the behavior of the malnourished
organism may be significantly altered by long-term changes in
motivation, attention span, and arousal without directly affecting
learning capacities or memory per se.

Rat studies have effectively shown that severe protein
malnutrition in infancy results in long-term behavioral changes,
including increased apathy and decreased exploratory and problem-
solving performance. These effects were most pronounced when
the interval of malnutrition combined both prenatal and infancy
periods. However, rats seem to be more sensitive to the effects of
protein restriction than primates. For example, rats do not survive
well on diets containing less than 10% protein, but monkeys survive
and grow adequately on many of these diets.

The large variance in the reaction of species to protein
deficient diets suggests that organisms differ in the degree to
which they can withstand the restriction. The severity of a diet,
therefore, may be defined in terms of the inability of the organism
to adapt or adjust physically, biochemically, or behaviorally to the
nutritional condition. It might be further assumed that many
primates, including humans, have evolved to show considerable
adaptability in response to nutritional adversity. By this reasoning,
within a certain range of mild to moderate malnutrition the
behavior of protein-restricted animals will reflect adaptive responses rather than "damage" to the organism. Severe diets are
severe because of the failure of adaptive mechanisms to compen-
sate for the restriction.

Our early experiments were designed to evaluate the effects
of short-term severe protein-calorie malnutrition in developing
rhesus monkeys between weaning and puberty. Differences were
found between protein-restricted animals and well-nourished controls, particularly in social interactions. However, these differences were minor compared with the extensive number of measures
that did not differentiate the groups. The malnourished animals
were impressive in their ability to remain relatively normal in the

face of nutritional adversity. It was only when malnutrition was combined with early social and environmental restrictions that the monkeys showed more severe signs of maladjustment.

These early studies provided impetus for further research to determine the effects of long-term nutritional deficiencies on behavioral development. Specifically, the question was raised whether protein restriction earlier in the so-called periods of brain vulnerability would produce a more devastating effect on behavior than later deprivations. Contrariwise, it was hypothesized that the earlier the restriction, the longer the animal would have to make compensatory responses to the treatment. Furthermore, the younger organism may well have more flexibility in adjustment in that neural mechanisms and behavioral patterns would not have been rigidly established.

Two studies are reported that measured the effects of prenatal malnutrition on social and behavioral development of rhesus macaques (Macaca mulatta).

PRENATAL ZINC DEFICIENCY

The essential importance of zinc in mammalian diets to maintain growth and body functioning has been well established (Hurley and Swenerton, 1966). Of particular interest is the possible damage produced in the nervous system by early zinc deficiencies (Hurley and Shrader, 1972). It is possible that some human teratology (Sever and Emanuel, 1972) and the occurrence of "small for date" infants might be related to zinc deficiencies during pregnancy (Metcoff, Mameesh, Jacobson, Costeloe, Crosby, Sandstead, and McClain, 1976).

Studies with rats support the idea that zinc nutrition is essential for normal development. When rats were deprived of zinc during the third trimester of pregnancy, abnormalities developed including decreased fetal weight, brain weight, and brain DNA (McKenzie, Fosmire, and Sandstead, 1975). Long-term effects of early and severe zinc restrictions in rats have been reported, including impaired avoidance performance as adults (Halas and Sandstead, 1975).

Unfortunately, the third trimester of pregnancy for a rat is relatively short (less than a week) and does not provide much time for the animal to make adaptive responses to the restriction. In the monkey the length of this period is more extensive, approximately 55 days for the rhesus macaque. A preliminary experiment, therefore, was established to produce zinc deficiency in pregnant rhesus monkeys during the period of rapid intra-uterine brain growth and to test their offspring for behavioral change.

Twelve females entering their third trimester of pregnancy were introduced to a 20% sprayed egg white, biotin-enriched, zinc-deficient diet that was originally developed for rats (Halas and Sandstead, 1975). All animals had been maintained on Purina Monkey Chow (Ralston-Purina) prior to that time. Four of these animals were fed the diet ad libitum (AL), but received 50 ppm zinc in their drinking water. Four animals were fed the diet without supplemental zinc in the water (ZD). The remaining monkeys were pair-fed (PF) with time-mated females in the ZD group and given 50 ppm zinc in their water. On the 150th day of gestation, 55 ppm zinc was added to the water of the ZD group. Following delivery, all groups were returned to the chow diet containing approximately 31 ppm zinc.

The zinc deprivation produced a significant drop in plasma zinc levels from a range of 46-76 µg% to a range of 8-23 µg%

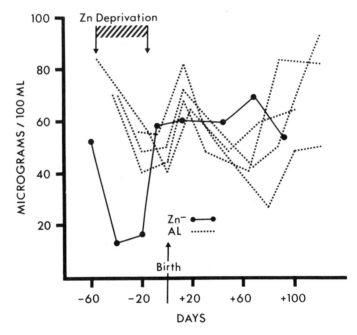

Figure 1.   Blood plasma zinc levels in mothers fed zinc-deficient diets with or without supplemental zinc during pregnancy.

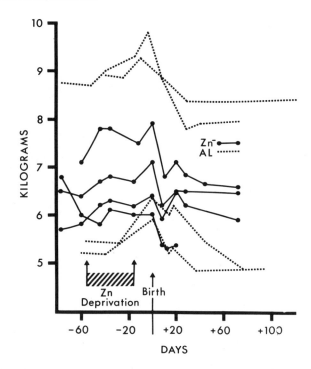

Figure 2.   Individual body weights of mothers that are zinc
deprived or zinc supplemented during pregnancy.

(Figure 1), and caused alopecia. In addition, a rash developed in
the genital area and the ventral surface of the females.

The pair-fed females showed little tolerance for the experimental manipulation. Two aborted their infants within the week
preceding expected normal delivery and one had an early neonatal
death following a breech delivery.

The AL infants had greater average birth weights (520µg)
than did ZD infants (505 µg), but the differences were not
statistically significant. The AL mothers showed a significantly
greater postparturition weight loss than the ZD mothers (Figure 2)
and the ZD infants showed a more rapid gain in weight than
controls (Figure 3).

No unusual differences were observed in mother-infant interactions during the next 6 months. During early periods of weaning,
between 2 and 4 months of age, the ZD infants spent significantly

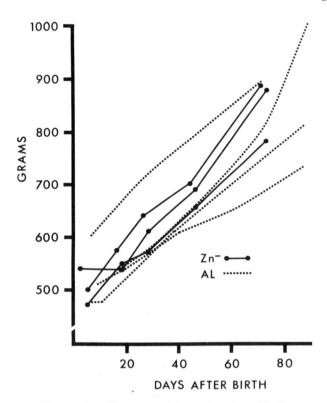

Figure 3.   Infant weight gain after birth.

more time than controls in associative behaviors with their
mothers (Table 1).   These associations involved more direct physical contacts and more nursing than displayed by AL monkeys. As a
result, the ZD infants played less, explored less, and were less
active than controls.

Following separation from their mothers at 6 months, peer-
peer interactions in the ZD group continued to appear normal
(Table 2).   The AL monkeys were more associative, however,
spending a greater proportion of their time engaged in ventral-
ventral clutching than ZD animals. When groups of animals were
mixed in a social playroom in "round-robin" fashion, few differences in behavior were noted (Table 3). The ZD animals showed a
significant decline in withdrawal behaviors and the AL monkeys
showed a tendency to shift more of their time from associative
behaviors to exploration.

TABLE 1. MOTHER-INFANT BEHAVIORAL INTERACTIONS IN ZINC DEFICIENT AND DIET CONTROLS

Measurement	Diet Group Means			Observations p value	Diet X Observations
	Zn-	AL	p value		
Association frequency	12.15	10.38	0.616	0.022	0.822
Play frequency	10.80	17.09	0.021	0.024	0.175
Exploration frequency	9.48	13.53	0.045	0.240	0.001
Association % time	59.62	45.95	0.044	0.253	0.065
Play % time	23.08	30.73	0.103	0.055	0.207
Exploration % time	15.80	19.52	0.157	0.001	0.033
Away from mother % time	30.08	39.61	0.160	0.819	0.174
Nursing % time	40.48	34.34	0.280	0.006	0.018
Activity level	2.23	2.44	--	0.415	0.101

TABLE 2. PEER-PEER SOCIAL INTERACTIONS ZINC STUDY-BASELINE

Behavior	Percent Duration			Frequency		
	Zn-	p value	AL	Zn-	p value	AL
Withdrawal	3.18	ns	1.38	0.49	ns	0.22
Exploration	31.98	ns	25.68	0.03	ns	0.22
Play	11.52	ns	7.18	0.07	ns	0.20
Assertive	0.625	ns	0.650	1.88	ns	1.70
Associative	32.45	0.008	45.70	15.99	ns	18.03
Non-Directive (passive)	20.68	ns	19.42	---	---	---

TABLE 3.   PEER-PEER SOCIAL INTERACTIONS ZINC STUDY-
            ROUND ROBIN

Behavior	Percent Duration			Frequency		
	Zn-	p value	AL	Zn-	p value	AL
Withdrawal	0.20	0.018	1.39	0.10	0.0015	1.09
Exploration	60.92	ns	74.96	3.24	ns	3.80
Play	0.63	ns	0.08	0.44	ns	0.09
Assertive	0.46	ns	0.08	0.26	ns	0.08
Associative	37.48	ns	22.93	2.49	ns	0.79
Non-directive (passive)	0.0	ns	0.0	0.0	ns	0.0

At 500 days of age, the ZD monkeys performd as well as controls on repeated color discrimination reversal problems using food rewards (Figure 4). However, at 554 days of age, the ZD monkeys were unable to show adequate interproblem transfer on 100 object-quality discrimination (Learning Set) problems (Figure 5). This failure to demonstrate Learning Set behavior in zinc-deprived monkeys may reflect a maturational delay in these animals.

CHRONIC PROTEIN MALNUTRITION

The second study differs from the first in that the period of malnutrition included not only the last period of intra-uterine life, but the period of infancy as well. Furthermore, the mothers of the subjcts in the study were reared from weaning to puberty on protein-deficient diets. Thus, the study was designed to examine the effects of chronic intergenerational effects of malnutrition on behavior.

Five protein-malnourished nulliparous female monkeys were rehabilitated after 1,290 days on diets containing 3.5% protein by weight. They were bred with experienced males and approximately 120 days from conception they were again placed on the protein-deficient diets. A control group of four pregnant females, previously reared on a commercial laboratory chow diet (15% protein), was given a diet containing 25% protein by weight.

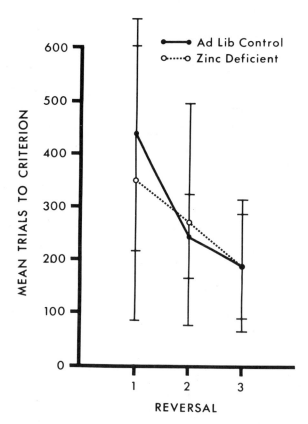

Figure 4.   Performance of zinc-deficient and control monkeys on repeated discrimination reversal problems.

One of the malnourished females had an unsuccessful breech delivery 28 days after receiving the low protein diet. No significant differences in birth weights were found between the diet groups.

Unlike the zinc-deprived mothers in the previous study, the protein-malnourished monkeys were less protective of their infants during the first 200 days of life. The low-protein infants spent significantly less time on their mothers than controls, and were often physically abused at feeding time.

All infants were weaned and separated from their mothers at 200 days of age and maintained on their respective diets. Social observations were made in a playroom between 550 and 800 days of age (Table 4). The patterns of peer-peer social interactions were similar to those found in previous observations of protein deficiency, but not nearly as severe. In general, the malnourished monkeys spent less time engaged in play behaviors, but proportionately more time exploring and engaging in contact behaviors, such as grooming and clutching, than controls. A series of observations was made with a "stranger" monkey present in the playroom. The stranger in each session was either a second-generation malnourished monkey or a dietary control monkey reared separately from the diet groups.

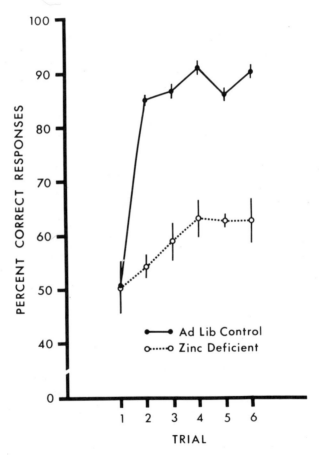

Figure 5. Performance of zinc-deficient and control monkeys on 100 object-quality learning set problems.

TABLE 4.   PERCENT DURATION OF MAJOR BEHAVIORS AND DIRECTION OF MAJOR BEHAVIORS IN SOCIAL PLAYROOM

Major Behavior	Baseline Observations Low Protein	High Protein	Stranger Observations Low Protein	High Protein	Round-Robin Observations Low Protein	High Protein
WITHDRAWAL	1.48 (ns)	0.68	1.97 (ns)	0.29	21.52 (p=.002)	0.10
Direction:						
withdrawing animal	—	—	16.22	3.61	—	—
exploring animal	6.16	0.0	18.46	0	73.375	0.0
playing animal	8.89	0.0	29.64	0	11.50	0
assertive animal	25.52	0.0	4.25	27.24	14.875	0
extraneous animal	53.18	100.00	31.44	69.15	0	100.00
other	—	—	—	—	0.375	0
EXPLORATION	76.68 (p=.04)	61.27	70.35 (ns)	58.63	76.62 (ns)	67.66
Direction:						
withdrawing animal	—	—	4.50	11.75	0	0.50
exploring animal	24.31	20.87	30.125	27.375	55.625	39.125
playing animal	5.73	30.00	7.875	17.125	0.25	0
objects	69.32	43.875	57.00	40.375	43.75	58.25
extraneous noise	0.04	5.125	—	3.37	0.375	2.125
other	—	—	.50	—	—	—
PLAY	15.00 (p=.04)	33.59	19.24 (ns)	34.31	1.21 (p=.0001)	17.48
Direction:						
withdrawing animal	—	—	0.50	2.25	0	1.875
exploring animal	—	—	0.67	0	0	1.50
playing animal	36.87	55.125	57.50	61.0	0.80	26.50
with self	49.37	42.875	40.50	35.875	99.20	69.75
with objects	1.33	1.875	0.83	1.00	0	0.375
ASSERTIVE	1.66 (ns)	0.43	1.83 (ns)	2.26	0.14 (ns)	2.18
ASSOCIATIVE	1.59 (p=.007)	0.43	3.14 (p=.004)	0.39	2.25 (ns)	3.26
NON-DIRECTIVE	1.65 (ns)	3.11	3.44 (ns)	4.05	2.25 (p=.03)	9.47

The well-nourished controls showed an initial increase in exploratory responses and concomitant decline in the proportion of play behaviors with the introduction of the stranger (Figures 6 and 7). However, these changes in behavior reversed with continued exposure to the unfamiliar animals. On the other hand, the low-protein monkeys reacted to the strangers by significantly increasing their grooming and systematically began to denude the high-protein stranger. Both groups showed an initial increase in assertive behaviors toward strangers of their own dietary condition, but this increase also declined with repeated exposures.

Further social tests were conducted by mixing the diet groups in round-robin pairings. The protein-malnourished monkeys showed

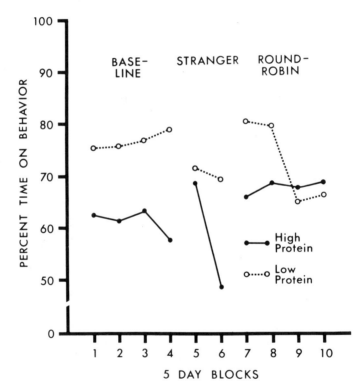

Figure 6.   Percent time on exploratory behaviors during peer-peer social interactions; second-generation protein malnutrition.

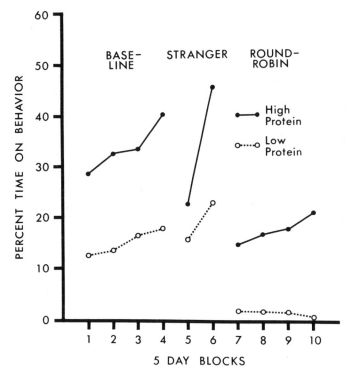

Figure 7. Percent time on play behaviors during peer-peer social interactions; second-generation protein malnutrition.

a significant increase in the duration and frequency of withdrawal responses to the high-protein monkeys (Figure 7) and significantly lower levels of play and nondirected behaviors than controls.

While social interactions, particularly play behaviors, seemed to have been disrupted by chronic protein malnutrition, there was little evidence for impaired learning performance in these animals. The only task that reliably produced differences in test performance as a result of protein insufficiency separated the location of color cues from the locus of response during discrimination reversal trials. Presumably, this manipulation made the task more difficult by placing greater demands upon attentional processes. The second generation protein-malnourished monkeys were significantly inferior to controls in preforming this task.

DISCUSSION

Within the framework of the conference on nursery care of nonhuman primates, the present studies dramatically indicate the resiliency of monkeys to prenatal nutritional insult, at least for M. mulatta. More radical disruptions in social behaviors have been found in monkeys deprived of protein after weaning (Zimmermann, Steere, Strobel, and Hom, 1972; Holombo, 1976) and in performance on attentional learning tasks (Zimmermann, Geist, and Wise, 1974).

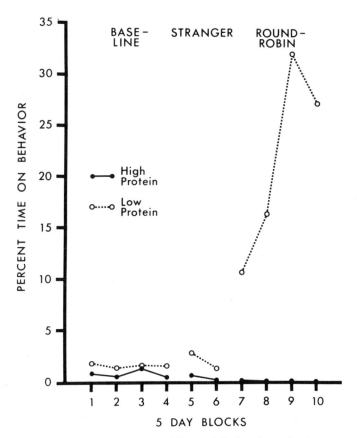

Figure 8.   Percent time on withdrawal behaviors during peer-peer social interactions; second generation protein malnutrition.

These results generally support the contention that early deprivations provide more time for the animal to adjust to the nutritional restriction. Both of the present studies provided sufficient socialization for the animals; each diet group was reared in a large community cage after weaning. Previous studies have found that early social restrictions combined with nutritional restrictions are necessary to produce radical changes in the behavior of the monkeys (Strobel and Munro, 1977), and that the environmental manipulation is a more salient variable than the nutritional variable.

The only monkeys that did not respond well to the experimental manipulations were the Pair Fed subjects of the zinc study. The high pregnancy risk in this group might be attributable to the stringent feeding schedule they experienced. Each pregnant female in this group was fed an amount of diet equivalent to the quantity consumed by a matched pregnant female in the zinc-deficient group. In rats, the pair-feeding group is necessary because the ZD subjects decrease their caloric intake over the deprivation interval, or show cyclic changes in the quantity of food ingested. On the average, the ZD monkeys consumed more of the diet than AL controls. However, they also showed the cyclic patterns observed in rats subjected to zinc deficient diets. Thus, there were days that the PF animals consumed as much as 247 g of the diet or as little as 28 g. The infants of the ZD females may have survived because they fluctuated their intake of food as part of some adaptive response to zinc restriction. By comparison the PF females had the caloric fluctuation imposed upon them, and they were unable to make an adaptive response so they aborted.

Indirect support for this hypothesis can be found in our recent pilot studies of iron deficiency anemia. In general, the monkeys showed the most radical disruption in their behavior during periods of dietary change, in deprivation and rehabilitation phases. This would suggest that possible prenatal effects of nutritional restriction may result in part from dietary change, as much as from the lack of essential nutrients. At least this possibility should be investigated before conclusive statements can be made about the behavioral effects of early malnutrition.

REFERENCES

Birch, H. G. and Gussow, J. Disadvantaged Children. Health, Nutrition, and School Failure, New York: Harcourt, Brace and World, Inc., 1970.

Cravioto, J. and DeLicardie, E. R.   Nutrition and behavior and
    learning. Wld Rec. Nutr. Diet 16:80-96, 1973.
Halas, E. S. and Sandstead, H. H.   Some effects of prenatal zinc
    deficiency on behavior of the adult rat. Ped. Res. 9:94-97,
    1975.
Holombo, L. K.   The effects of differential diets and enviornments
    on the social behavior of monkeys.   Unpublished Master's
    Thesis, University of Montana, 1976.
Hurley, L. S. and Shrader, R. E.   Congenital malformations of the
    nervous system in zinc deficient rats.   Pp. 7-51 in: Interna-
    tional Review of Neurobiology, Supplement 1, C. C. Pfeiffer
    (Ed.), New York, Academic Press, 1972.
Hurley, L. S. and Swenerton, H.   Congenital malformation resulting
    from zinc deficiency in rats.   Proc. Soc. exp. Biol. Med.
    123:692-696, 1966.
Latham, M. C.   Protein-calorie malnutrition in children and its
    relation to pychological development and behavior.   Physiol.
    Rev. 53:541-565, 1974.
McKenzie, J. M., Fosmire, G. J., and Sandstead, H. H.   Zinc
    deficiency during the latter third of pregnancy: Effects on
    fetal rat brain, liver, and placenta. J. Nutr. 105:1466-1475,
    1975.
Metcoff, J., Mameesh, M., Jacobson, G., Costeloe, P., Crosby, W.,
    Sandstead, H., and McClain, P.   Maternal nutrition and
    leukocyte metabolism at mid-pregnancy related to baby/
    placenta size at term. Clin. Res. 24:502A, 1976.
Read, M. S.   Malnutrition, hunger, and behavior. I. Malnutrition
    and learning. J. Amer. diet. Ass. 63:379-385, 1973.
Sever, L. E. and Emanuel, I.   Is there a connection between
    maternal zinc deficiency and congenital malformation of the
    central nervous system in man? Teratology 7:117-118, 1972.
Strobel, D. A. and Munro, N.   Behavioral changes in malnourished
    monkeys.   In: Handbook of Nutrition and Food, M. Rechcigl,
    Jr. (Ed.), Cleveland: CRC Press, in press, 1977.
Zimmermann, R. R., Geist, C. R., and Wise, L. A. Behavioral
    development, environmental deprivation, and malnutrition.
    Pp. 133-192 in: Advances in Psychobiology, Vol. 2, G.
    Newton and A. H. Reisen (Eds.), New York: John Wiley and
    Sons, Inc., 1974.
Zimmermann, R. R., Steere, P. L., Strobel, D. A., and Hom, H. L.
    Abnormal social development of protein-malnourished rhesus
    monkeys. J. abnorm. Psychol. 80:125-131, 1972.

II. EARLY ASSESSMENT

CHAPTER 6

AGE DETERMINANTS IN NEONATAL PRIMATES:
A COMPARISON OF GROWTH FACTORS

M. Michejda and W. T. Watson

Primate Research Unit
Veterinary Resources Branch
Division of Research Services
National Institutes of Health
Bethesda, Maryland

INTRODUCTION

Interest in nonhuman primates as animal models for biomedical research can be traced back four decades, when rhesus macaques (Macaca mulatta) were used for the first time at Yale University in studies of developmental and reproductive physiology (Hartman, 1932; van Wagenen, 1972). During many years of using nonhuman primates in various areas of biomedical research, an alarming decrease in the wild population of various species has been observed. Thus, the increasing demand for nonhuman primates resulted in the development of large breeding programs and increasing emphasis on domestic breeding rather than importation of various species. However, this requires more information on neonatal care as well as on the early developmental stages of these species in order to raise healthy primates in captivity. The purpose of our study was to contribute to the better understanding of various problems related to the neonatal stage of development in nonhuman primates by obtaining information on growth and development as well as age characteristics and age determination of neonatal M. mulatta. Further, we sought to establish criteria for the best determination of age and its related growth changes in neonates and infants, and to correlate certain growth characteristics with age. Finally, we compared various methods of age determination commonly used in biomedical research to determine which of these is best for the period from birth to 6 months of age.

MATERIALS AND METHODS

Longitudinal studies are being conducted on 15 infant rhesus macaques of both sexes from birth through 6 months of age. However, this chapter will deal only with the first 3 months of life. The infants, born at the Primate Research Unit of the National Insitutes of Health, were the offspring of timed pregnancies, and the gestational age of each newborn monkey was accurately established within ±48 hours. Six infants were derived from a prenatal longitudinal study of appendicular bone maturation, in which multiple uterotomies were performed. Comparison was made of gestational age and body weight between the infants that experienced surgical invasion in utero and the rest of the population.

Owing to the small sample and insignificant sexual dimorphism at the early age, no sex subgrouping was applied. The infants were kept with their mothers in individual stainless steel cages, in a uniform environment. The nursing females were fed a standard laboratory diet with additional fruits and water ad libitum. On the first day of postnatal life, each neonate was radiographed, measured, and weighed. Each infant was examined and data were obtained at weekly intervals from birth to 8 weeks, biweekly intervals from 8 to 12 weeks, and at monthly intervals from 12 to 24 weeks (Watts, 1977). Infants over 4-6 weeks old were tranquilized for the convenience of examination (Vetalar(R), Parke-Davis and Co., Detroit, MI). The following growth dimensions were measured: head length and breadth; sitting height; humerus, radius, femur, and tibia lengths; body weight; tooth eruption (dental age); and carpal bone ossification (skeletal age). All anthropometric measurements were obtained by the same observer (M.M.) and the error in the body measurements attributable to inconsistency in techniques was avoided by a three-fold repetition of each measurement for each growth parameter. Linear measurements were recorded in millimeters using a sliding caliper, and standard anthropometric techniques were applied (Schultz, 1929). The body weight was recorded to the nearest gram, and tooth eruption was recorded and scored when the tooth crown penetrated the gingiva and was well visible. The assessment of skeletal age was attained from roentgenograms of the left hand and wrist. The radiographic techniques applied are listed in Table 1. The quantitative assessment and scoring of the ossification centers in the hand and wrist were based on the methods of Greulich and Pyle (1959) and compared with data obtained by van Wagenen and Asling (1958). The means were compared using Student's t-test.

TABLE 1. RADIOGRAPHIC TECHNIQUES

FILM:	DuPont Cronex 4: Size 5 x 7
CASSETTE:	Picker Kromsteel 5 x 7 with Intensifying Screen
SCREEN:	DuPont Cronex Hi-Plus
X-RAY:	Apparatus: Westinghouse (1500 mA) model "Delray"
DEVELOPING MACHINES:	Kodak RP X-Omat Processor
EXPOSURE SETTING:	40 KV; 300 mA
EXPOSURE TIME:	1/30 sec
CONSTANT TUBE TO TARGET DISTANCE:	40 in

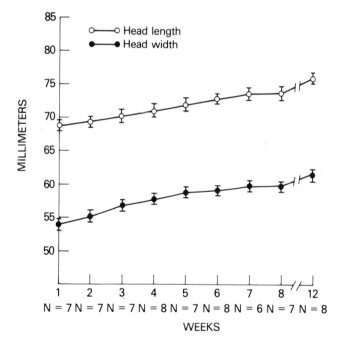

Figure 1.   M. mulatta head length and width (mean ±1 S.E.).

RESULTS AND DISCUSSION

Cranium

Results of cranial measurements are summarized in Figure 1. Development of the cranium proceeds in a linear fashion with little variation within groups (S.E. <1 mm). Although the variation within groups is small, the overlap between groups (i.e., 1 week and 2 weeks, etc.) precludes using these parameters to differentiate between infants whose ages differ by 1 week or more. Cranial measurements, however, may be of value in setting limits. For example, a rhesus monkey with a head length of 71 mm and a width of 58 mm would be between 2 and 8 weeks of age. These measurements may also be of value when used in conjunction with other factors to assess age in nonhuman primates (Sirianni, Swindler, and Tarrant, 1975; Ferron, Miller, and McNulty, 1976). Data from additional subjects within the age group under study are needed to determine whether sex differences exist.

Sitting Height

The relation between age and sitting height is shown in Figure 2. These data compare favorably with those presented by others (Gavan and Swindler, 1968; Gavan and Hutchinson, 1973). It appears that the height of rhesus monkeys increases 4-6 mm per week through 12 weeks of age. This is not surprising when one assumes that the increase in the length of the axial skeleton is proprotional to the overall growth and development of other organs in young animals. Based on the data presented, one could assess the age of rhesus neonates at 4-week intervals with some degree of confidence since deviations do not overlap. Sitting height may be used to monitor the normal growth and development of neonatal nonhuman primates as it is used in human neonates.

Appendicular Skeletal Growth

Long bone development follows a pattern similar to that of cranial development in that it is linear (Figure 3) with little variation within groups (S.E. ≤1.75 mm in most groups). However, it is of little value in distinguishing between neonates less than 4 weeks apart. These values, when taken collectively and in combination with other factors described, may allow for placement of neonates in broad age categories.

Body Weight

Figure 4 shows that body weights of rhesus macaques vary substantially in each age group. Body weight is the least accurate method of age assessment of all the factors studied, although it is the most practical to determine. Some of the observed variability

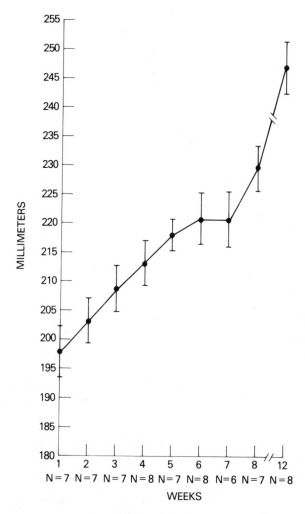

Figure 2.   M. mulatta sitting height (±1 S.E.).

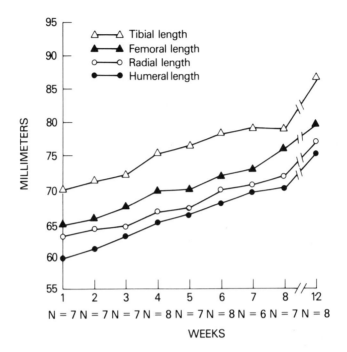

Figure 3.   Mean appendicular skeletal growth in M. mulatta.

might be reduced if each sex were grouped separately, although Haigh and Scott (1965) noted that the weights of males and females at 6 months of age were similar. Many factors contribute to the wide variation in body weight of neonates, including quantity and quality of milk produced by the mother, quality of diet fed to the mother, environment stresses in the colony during lactation, temporary illness leading to anorexia, and individual feeding habits of the neonate. Although body weight would not be an accurate method for critical age assessment, it may be of value in placing monkeys in broad age categories if their previous histories are known.   Body weights of colony-born, mother-reared neonates differ from those of colony-born, nursery-reared neonates since conditions are better controlled in the latter instance (Zimmerman, 1969; van Wagenen, 1972; Southam, 1973).

Dental Age

The dental age expressed by the sequence of eruption of the deciduous and permanent teeth of nonhuman primates for many years has been considered to be an important age determinant (Schultz, 1933; Nissen and Riesen, 1949, 1964; Eckstein, 1948; Hurme and van Wagenen, 1953, 1956, 1961; Voors and Metselaar,

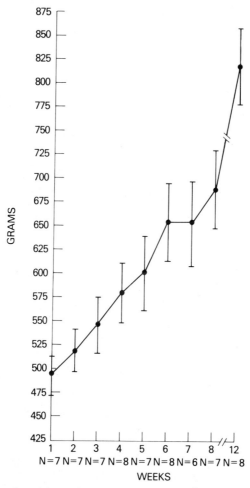

Figure 4.   M. mulatta body weights (mean ± 1 S.E.).

1958; Ockerse, 1959; Hurme, 1960; Berkson, 1968; Kenney, 1975).
Dental age has been used as a valid parameter and an integral part
of many growth studies (Rahlmann and Pace, 1969; Bowen and
Koch, 1970; Gavan et al., 1973).   However, the long intervals
between eruptions of various groups of dentition (6-10 months)
make such age assessment incomplete. Moreover, in our experience, the dental age of animals born in our colony was much more
accelerated when compared with Schultz's standards, possibly
owing to improved nutrition as well as early weaning.   Growth
factors such as body weight and dental age used routinely for age
estimation are too dependent upon extrinsic, intrinsic, or health
factors and have proven to be too variable.   Thus, better nutritional conditions for laboratory-born animals, particularly early-
weaned infants, may contribute to significant increases in body
weight and earlier eruption of teeth than in animals raised in
natural or seminatural habitats.

The eruption of deciduous teeth in our population of rhesus
infants was accelerated by approximately 2 weeks when compared
with dental age standards established by Schultz (1929, 1930). The
sequence of eruption was the same as for all other primates, with
the following pattern: central lower incisors followed by upper
central incisors, lower lateral incisors, upper lateral incisors,
canines, and molars.   The eruption of lower and upper central
incisors in our population was completed in the fourth week of
postnatal life, while Schultz's standards indicated the sixth week.
Lower and upper lateral incisors were all present in our sample at 6
weeks and erupted between 8 and 10 weeks, while Schultz's
standards indicate 12-14 weeks. First deciduous molars were fully
erupted in our population during 12 weeks of postnatal life while
Schultz's standards indicate 14 weeks.   The dental age data are
summarized and tabulated in Table 2. Owing to the small sample

TABLE 2.   TOOTH   DEVELOPMENT   IN   RHESUS   MONKEYS
(FEMALES AND MALES)

Week	No. animals	Range	Mean	S.D.
2	3	2-3	2.3	0.5
4	7	4-6	4.9	1.0
6	8	6-8	7.8	0.66
8	8	8-12	9.2	1.7
10	7	8-14	11.4	2.3
12	5	12-16	15.2	1.9

and nonsignificant sex variations, there is no sex subgrouping. The infants belonging to prenatal study with multiple surgical invasions were excluded from the statistical evaluation. The dental age data followed closely the general physical and skeletal development of each infant.

Skeletal Age

The pioneering work of van Wagenen and her associates in the late fifties, and more recent studies carried out on nonhuman primates (Gisler, Wilson, and Hekhius, 1962; Haigh et al., 1965; Bramblett, 1969; Dobbelaar and Arts, 1972; Watts, 1975; Phillips, 1976; Newell-Morris, personal communications) have indicated that the best criterion for age estimation is appendicular bone maturation. Moreover, as mentioned before, such parameters as body weight and dental age used routinely for age estimation are too dependent on extrinsic, intrinsic, or health factors and have proven to be arbitrary. The ossification and fusion of carpal and meta-carpal bones are among the most accurate criteria for assessing the age of nonhuman material, and have been used for years in pediatrics and forensic medicine (Greulich et al., 1959) as well as in primatology. However, the acceleration of bone maturation in nonhuman primates is signficant when compared with that in man. In newborn M. mulatta the bone ossification resembles that of a 5- to 8-year-old child (Noback, 1954) or a 15- to 18-month old chimpanzee (Nissen and Riesen, 1949). The onset of wrist ossification in M. mulatta is expected at about 120 days of gestational age. This stage of bone development corresponds to that in a newborn child. The average age of onset of ossification in epiphyseal and short bone centers is 12-20 months earlier in the chimpanzee than in man (Niesen et al., 1949; Gavan, 1953; Keeling and Riddle, 1975). It seems that in higher primates the relative placement on the evolutionary scale is inversely proportional to the ossification and bone maturation. In spite of this significant acceleration of growth, great similarities in the mode of bone maturation and sequence of ossification in the macaque hand and human hand and wrist were found (Hill, 1939; Schultz, 1956; Washburn, 1943; Randall, 1943) and can be used in comparative studies.

Radiographic evaluation of skeletal age obtained at weekly intervals indicated slower postnatal growth changes when com-pared with prenatal development. Moreover, little or no sexual dimorphism was observed during the first 3 months of postnatal life and all measured growth parameters seemed to be strictly a function of gestational age at this early stage of postnatal develop-ment.

No significant growth changes were recorded at weekly intervals; instead, typical growth changes and scoring of ossification centers observed at biweekly intervals are discussed.

In our newborn M. mulatta (Figure 5) the typical radiographic picture of bone maturation corresponded to earlier findings (van Wagenen et al., 1958). Six to seven carpal ossification centers (scaphoid, trapezium, capitate, hamate, triquetral, lunate, and pisiform and radial epiphyses) were present in both sexes. However, the ulnar epiphysis was not constant and was observed only in 10% of our population. No consistent sex variations were observed in this area. Epiphyses of four to five metacarpals were also present. In newborns of 158-162 days gestation the seven existing ossification centers showed lower density and overall size than those of

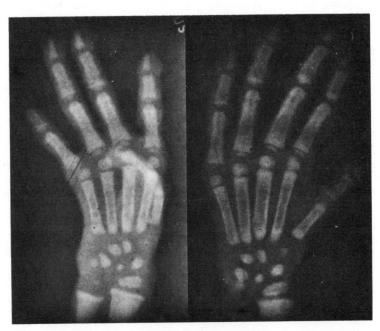

Figure 5.   Radiographs of two M. mulatta newborns of the same gestational age (160 days) showing the same number (7) of carpal ossification centers. Left, female; right, male. The overall size and density of the carpal ossification centers, metacarpal epiphyses, distal ulnar and radial epiphyses and phalanges are similar. This state of development of the hand and wrist is almost identical in both sexes.

neonates born after 162 gestational days. Thus, the stage of bone maturation in newborns did not show any consistent sexual dimorphism, but was more related to the gestational age. The average stage of skeletal development of the wrist and hand in our population of newborns corresponded to that of a 5½-year-old child.

At 2 weeks of postnatal life, seven carpal bones were observed consistently, indicating that pisiform ossification centers developed in the first 2 weeks of postnatal life. Moreover, the growth changes in the area were limited to the significant increase in size of carpal bones, not the number of ossification centers. All four proximal phalangeal epiphyses increased in overall size, but no fusion was observed. This was also true for the distal radial and ulnar epiphyses. The latter was present in all specimens at that age. In addition, the distal radial epiphysis started to attain the wedge-like shape.

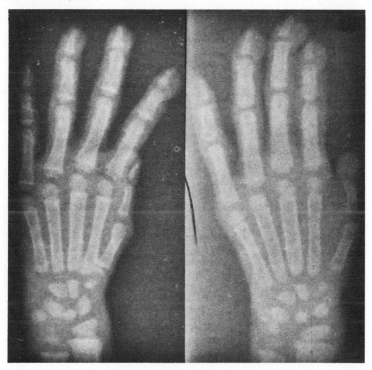

Figure 6. The same animals as in Figure 5, shown at the age of 4 weeks. All the morphological characteristics cited in Figure 5 are still the same for both sexes, and no evidence for sexual dimorphism is observed at this age.

At 4 weeks of postnatal life (Figure 6) the same seven carpal ossification centers were observed. The overall size and density of these centers were comparable in both sexes. However, the distal ulnar epiphysis was more developed in females than in males. All four proximal phalangeal epiphyses were well developed and the onset of ossification of the fifth was observed only in females, which may represent the first indication of sexual dimorphism. The skeletal age of infant females in our population corresponded to that of a 5.25-year-old girl and a 5.5-year-old boy.

At 6 weeks, the same seven ossification centers were observed in both sexes. However, the carpal bones were larger in

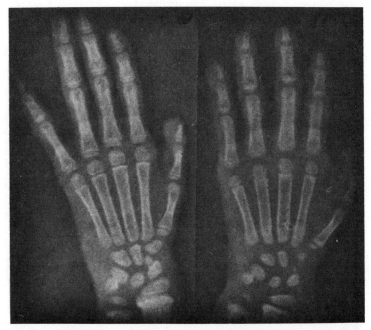

Figure 7.   The same M. mulatta as in Figures 5 and 6, at the age of 2 months.   Eight carpal ossification centers are present in both sexes. However, the size and stage of ossification of carpal bones is more advanced in the female infant (left). The maturation of the distal ulnar epiphyses is much more advanced in the female when compared to that in the male. Epiphyses of 5 meta-carpals are present. However, the proximal phalangeal epiphysis of the first digit is present only in the female monkey, indicating the first sign of sexual dimorphism.

females than in males, and the pisiform ossification center showed more density. This was also true for the radial and ulnar epiphyses, which in the males were better defined radiographically than they had been at 4 weeks. Four metacarpal epiphyses were present, but the onset of ossification of the fifth was still visible only in the females.

At 8 weeks of postnatal life (Figure 7) the same seven carpal bones were observed in males, and the overall size and the stage of bone maturation were comparable to those observed at 6 weeks. This was also true for all the phalangeal and metacarpal epiphyses. However, the onset of ossification of the fifth metacarpal epiphysis was observed. In female infants few growth changes were noted. In 16% of our female population, the eighth carpal ossification center (trapezoid) was found. Those females were born

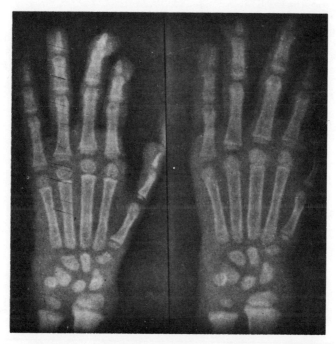

Figure 8.   The same M. mulatta infants of Figure 5 at the age of 3 months. The overall size of carpal bones is similar in both sexes. However, the stage of ossification is more advanced in the female (left) and the proximal phalangeal epiphysis of the first digit is still not visible in the male monkey (right).

after 162 days gestation, so it seems that the gestational age was closely correlated with skeletal age.

At 10 weeks of postnatal life, no significant growth changes were observed in the male population. In females, however, the onset of the ossification of the os centrale was observed.

At 12 weeks (Figure 8) of postnatal life, the last carpal ossification center (os centrale) was visible in males. Thus, this age indicator developed 2 weeks later than that of females. The ossification centers of the proximal phalanges of the first digit were present in all males. In females, the last maturity indicator among carpal ossification centers (os centrale) was well defined and no new centers were observed. It seems that the growth had stabilized at that age and slowed significantly in both sexes when compared with that of newborns and infants during their first 2 weeks of postnatal life.

Thus, it would appear that there is a possible relation between gestational age and such parameters as skeletal age (Figures 5-8) and birth weight, although no significant differences could be demonstrated between sexes (Tables 3 and 4). It was

TABLE 3. BIRTH WEIGHTS (GRAMS) OF Macaca mulatta

	Range	Mean	S.D.
Female (N=5)	395-540	464.0	62.0
Male (N=4)	390-500	451.3*	46.6

*No significant differences (p > 0.05)

TABLE 4. GESTATION AND DELIVERY (DAYS) FOR M. mulatta

	Range	Mean	S.D.
Female (N=6)	153-172	163.8*	8.33
Male (N=6)	148-167	158.5*	6.41

*No significant differences (p > 0.05)

noted during this study that the monkeys of longer gestation were heavier and had more advanced skeletal development. Most of the infants with longer gestations were females and a larger number of subjects are needed to prove statistical significance between gestation periods, birth weights, and sex.

SUMMARY AND CONCLUSIONS

1. Cranial measurements have only limited value as age indicators in M. mulatta. This is also true of other parameters, such as sitting height and the length of the long bones.

2. Body weight is the least accurate method of age determination.

3. Since the intervals between eruption of teeth are long, dentition is a poor indicator of age in young animals. The data also show that the eruption of the deciduous teeth was accelerated by approximately 2 weeks relative to Schultz's standards.

4. The most accurate age indicator is skeletal age. The ossification and maturation of the hand and wrist are most rapid during the first 2 weeks of postnatal life. Moreover, no significant sexual demorphism was observed at that early stage of postnatal development.

5. The first radiographically detected signs of sexual dimorphism were observed during the fourth week of postnatal life.

6. The skeletal age was closely correlated with gestational age.

ACKNOWLEDGMENTS

This research was funded, in part, by the Division of Research Resources and the Division of Research Services, National Institutes of Health. Animals used in this study were housed in a facility fully accredited by the American Association for Accreditation of Laboratory Animal Care.

We thank Dr. David K. Johnson for his support and encouragement, Dr. James Ferguson for his contribution to the statistical analysis of the data, Mr. George Coleman and the staff of the Primate Research Unit for their technical assistance, and Mrs. Dora Olson and Mrs. Norma Anousheh for their expert clerical assistance.

REFERENCES

Berkson, G. Weight and tooth development during the first year in Macaca irus. Lab. anim. Care 18:352-355, 1968.

Bowen, W. H. and Koch, G. Determination of age in monkeys (Macaca irus) on the basis of dental development. Lab. anim. Care 4:113-123, 1970.

Bramblett, C. A. Non-metric skeletal age changes in the Darajani Baboon. Amer. J. phys. Anthropol. 30:161-171, 1969.

Dobbelaar, N. Y. and Arts, T. H. M. Estimation of the foetal age in pregnant rhesus monkeys (Macaca mulatta) by radiography. Lab. anim. Care 6:235-240, 1972.

Eckstein, F. M. P. Age changes in dentition in the rhesus monkey. Brit. med. J. 2:168, 1948.

Ferron, R. R., Miller, R. S. and McNulty, W. P. Estimation of fetal age and weight from radiographic skull diameters in rhesus monkey. J. med. Primat. 5:41-48, 1976.

Gavan, J. A. Growth and development of the chimpanzee: A longitudinal and comparative study. Hum. Biol. 25:93-143, 1953.

Gavan, J. A. and Swindler, D. R. Growth roles and phylogeny in primates. Amer. J. phys. Anthropol. 27:181-190, 1968.

Gavan, J. A. and Hutchinson, T. C. The problem of age estimation: A study using rhesus monkeys (Macaca mulatta). Amer. J. phys. Anthropol. 38:69-82, 1973.

Gisler, D. B., Wilson, S. G. and Hekhius, G. L. Correlation of skeletal growth and epiphyseal ossification with age of monkeys. Ann. NY Acad. Sci. 85:64-66, 1962.

Greulich, W. W. and Pyle, S. J. Radiographic Atlas of Skeletal Development of the Hand and Wrist. Stanford Univ. Press, 1959.

Haigh, M. V. and Scott, A. Some radiological and other factors for assessing age in the rhesus monkey using animals of known age. Lab. anim. Care 15:57-73, 1965.

Hartman, G. Studies in the reproduction of the monkey Macacus (pithecus) rhesus, with special reference to the menstruation and pregnancy. Contrib. Embryol. Carneg. Inst. 134, 23:1-161, 1932.

Hill, A. H. Fetal age assessment by centers of ossification. Amer. J. phys. Anthropol. 24:251-272, 1939.

Hurme, V. O. Estimation of monkey age by dental formula. Ann. NY Acad. Sci. 85:795-799, 1960.

Hurme, V. O. and van Wagenen, G. Basic data on the emergence of deciduous teeth in the monkey (Macaca mulatta). Proc. Amer. Phil. Soc. 97:291-315, 1953.

Hurme, V. O. and van Wagenen, G. Emergence of permanent first molars in the monkey (Macaca mulatta), association with other growth phenomena. Yale J. Biol. Med. 28:538-567, 1956.

Hurme, V. O. and van Wagenen, G. Basic data on the emergence of permanent teeth in the rhesus monkey (Macaca mulatta). Proc. Amer. Phil. Soc. 105:105-140, 1961.

Keeling, M. E. and Riddle, K. E. Reproductive, gestational and newborn physiology of the chimpanzee. Lab. anim. Sci. 25:822-828, 1975.

Kenney, E. B. Development and eruption of teeth. Pp. 154-167 in: The Rhesus Monkey, Vol. I, G. H. Bourne (Ed.), New York: Academic Press, 1975.

Nissen, H. W. and Riesen, A. H. Onset of ossification in the epiphyses and short bones of extremities in chimpanzees. Growth 13:45-70, 1949.

Nissen, H. W. and Riesen, A. H. The eruption of the permanent dentition of chimpanzee. Amer. J. phys. Anthropol. 22:285-294, 1964.

Noback, C. R. The appearance of ossification centers and the fusion of bones. Amer. J. phys. Anthropol. 12:63-69, 1954.

Phillips, T. R. Skeletal development in the fetal and neonatal marmoset (Callithrix jacchus). Lab. anim. Sci. 10:317-333, 1976.

Ockerse, T. The eruption sequence and eruption time of the teeth of the vervet monkey. J. dent. Ass. Sth Afr. 14:422-424, 1959.

Rahlmann, D. F. and Pace, N. Anthropoidimetric and roentgenographic growth changes in young pig-tailed monkeys (Macaca nemestrina). Proc. 3rd int. Cong. Primat. 2:171-180, 1969.

Randall, F. E. The skeletal and dental development and variability of the gorilla. Hum. Biol. 15:236-242, 1943.

Schultz, A. H. The technique of measuring the outer body of human fetuses and primates in general. Contrib. Embryol. Carneg. Inst. 213-257, 1929.

Schultz, A. H. Growth and development. In: Fetal growth and development of rhesus monkey, C. G. Hartman and W. L. Strauss, Jr. (Eds.), Baltimore: Williams and Wilkins Co., 1933.

Schultz, A. H. Postembryonic age changes. Primatologica 1:887-964, 1956.

Sirianni, J. E., Swindler, D. R., and Tarrant, L. H. Somatometry of newborn Macaca nemestrina. Folia primat. 24:16-23, 1975.

Southam, L. Comparison of two early separation and weaning schedules of infant rhesus monkeys (Macaca mulatta). J. med. Primat. 2:302-307, 1973.

van Wagenen, G. Vital statistics from breeding colony. J. med.
    Primat. 1:3-28, 1972.
van Wagenen, G. and Asling, C. W. Roentgenographic estimation
    of bone age in the rhesus monkey (Macaca mulatta). Amer.
    J. Anat. 103:163-184, 1958.
Voors, A. W. and Metselaar, D. The reliability of dental age as a
    yard stick to assess the unknown calendar age. Trop. geogr.
    Med. 10:175-180, 1958.
Washburn, S. L. The sequence of epiphyseal union in Old World
    monkey. Amer. J. Anat. 72:339-360, 1943.
Watts, E. S. The assessment of skeletal development in the rhesus
    monkey (Macaca mulatta) and its relationship to growth and
    sexual maturity. Pp. 245-259 in: The Rhesus Monkey, Vol.
    2, G. Bourne (Ed.), New York: Academic Press, 1975.
Watts, E. S. Some guidelines for collection and reporting of
    nonhuman primate growth data. Lab. anim. Sci. 27:85-89,
    1977.
Zimmerman, R. R. Early weaning and weight gain in infant rhesus
    monkeys. Lab. anim. Care 19:644-647, 1969.

CHAPTER 7

ASSESSMENT OF SKELETAL GROWTH AND MATURATION OF PREMATURE AND TERM Macaca nemestrina

C. E. Fahrenbruch, T. M. Burbacher, and
G. P. Sackett

Infant Primate Research Laboratory of the
Regional Primate Research Center and
Child Development and Mental Retardation Center
University of Washington
Seattle, Washington

INTRODUCTION

Primate nurseries have traditionally collected weight and intake data to construct growth norms for their populations. The purpose of this paper is to suggest supplementing these data with skeletal maturation and linear body growth information. Methods of assessment used at the Infant Primate Research Laboratory at the University of Washington are presented, using as an example a current study which contrasts the physical growth and ossification status of premature and term-gestation pigtail macaques (Macaca nemestrina). The postnatal growth and development of the macaque is of special interest as a model for the human infant. The fetal and infant growth of the macaque is similar to the human skeletal growth and maturation process (Kerr, 1975; Newell-Morris and Tarrant, 1978). Hand and foot ossification is much more rapid than in the human, reducing follow-up time for the appearance of round bones and epiphyses to months instead of years of postnatal age. Previous studies have indicated the value of ossification onset and increments as measures of skeletal maturity (Pryor, 1906; Flory, 1936; Garn and Rohmann, 1959, 1968; van Wagenen and Asling, 1964; Kerr, 1975). Skeletal growth and ossification assessment can provide a more detailed picture of infant maturation status than age estimates and weight alone and may have predictive value for later growth and development.

SUBJECT CHARACTERISTICS

Twenty-one M. nemestrina infants of known age were classified by gestational length as "premature" or "term." The subjects
were drawn from a larger sample of 51 infants of known age with a
mean gestation of 170 ±8.5 days. Five infants with gestations less
than 162 days were classified as premature. The control group was
composed of 10 males and 6 females ranging in age at birth from
162 to 179 days post-conception. The 21 infants were available for
follow-up from two studies. Eighteen subjects were from a study of
pregnancy outcome, maternal risk, and environmental stress (Chapter 2). Three of these infants were premature. The remaining
three infants were from a study of Idiopathic Respiratory Distress
Syndrome (IRDS). None of the premature subjects showed evidence
of the syndrome. Analysis indicated no differences between
subjects' mothers assigned to the two studies in parity (all were at
least parity three), and weight before the index pregnancy. The
IRDS-study mothers were more likely to have had a prior Caesarean section, dictated by research assignment rather than clinical
necessity.

METHODS

The subjects were assessed for physical size and skeletal
maturation at birth. Follow-up assessments were taken at standard
days of age from conception: 173, 187, 215, 243, 271, 299, and 355.
The schedule was designed for more frequent assessment during the
neonatal period, when postnatal acquisition of hand and foot
ossification centers is most rapid. Anthropometric measures taken
included crown-rump length and left foot length. A sliding scale
marked in millimeters was constructed to standardize these measures. For crown-rump length the infant was positioned with spine
flat against the scale, rump against the endplate, shoulders held so
that a line drawn between them was perpendicular to the scale, and
head positioned with the infant looking straight up. The sliding
endplate was moved against the crown, the infant removed, and the
measurement read from the scale. Left foot length was defined as
the length from the heel to the tip of the third digit. The heel was
held against the endplate with the plantar surface flat against the
scale. Two independent measures were taken and the mean value
used for analysis. Hand and foot radiographs were taken at a 1 m
distance at 32 kv, 200 ma, and 1/30 sec.

Each of the 30 ossification centers in the hand (carpals,
digital epiphyses, and distal radial and ulnar epiphyses) and 28
centers in the foot (tarsals, digital epiphyses, and distal tibia and

fibula) was scored as 0, 1, or 2. These values were assigned to differentiate among centers not yet ossified (0), centers in the initial stage of ossification (1), as described by Tanner et al. (1972), and centers past the initial ossification stage (2). The number of centers ossified was also recorded. The values assigned to the individual bones were summed for a maturity score. Garn and Rohmann (1959) reported that the metacarpal and digital epiphyses carry somewhat more maturation information than the carpals, but found the difference to be slight. Thus, each center is weighted equally in our scoring system. Sixty points were possible for the hand and 56 points for the foot. Maturity scores were assigned to the individual hand and foot ossification centers by two persons. Analysis indicated 98% reliability between them.

Using age from conception and age squared as the independent variables, multiple regression analysis of the control group data predicted maturation or growth rates for the four dependent variables: hand maturation, foot maturation, crown-rump length, and foot length. Each subject's mean deviation score from the control group predicted regression equation was calculated for the four variables by summing the deviations from regression at each assessment age and dividing by the number of assessments.

TABLE 1. CHARACTERISTICS OF M. nemestrina CONTROL GROUP AT BIRTH

	Males		Females	
	Mean	S.D.	Mean	S.D.
Hand total	25.90	2.42	24.17	2.04
Hand maturation	48.30	5.12	45.17	4.12
Foot total	24.33	1.41	22.67	2.73
Foot maturation	46.0	3.16	43.0	4.47
Foot length (mm) (p* <.01)	84.11	1.90	78.25	4.50
Crown-rump length(mm) (p <.05)	207.11	6.70	196.0	9.03
Birth weight (g) (p <.001)	610.20	47.12	488.33	43.09
Gestation (days) (p <.05)	172.0	4.11	166.83	4.54

* t-test

RESULTS

Birth Data

Tables 1 and 2 summarize the birth data for the term- and short-gestation groups. Control group birth assessments indicated no sex differences in the number of centers ossified or the hand and foot maturation scores. Newborn males, however, had longer foot and crown-rump lengths. One control group female was small for date compared with others of the same gestation. The other infants had normal sex-specific birth weight (unpublished data). One of the premature infants was not assessed at birth and another had only hand data. The male delivered at day 161 was small for date with a birth weight of 390 g. The 144-day female was heavier than expected for age and sex (Newell-Morris and Tarrant, personal communication).

TABLE 2.   BIRTH CHARACTERISTICS OF THE PREMATURE
M. nemestrina INFANTS

Subject	Sex	Delivery Type	Hand Maturation	Hand Total	Foot Maturation	Foot Total
FD	F	Surgical	32	17	21	13
II	F	Surgical	13	12	6	4
IV	F	Surgical	30	16	16	9
Y2	M	Natural	24	15	--	--
X6	M	Natural	--	--	--	--

Subject	Sex	Delivery Type	Foot Length	Crown-Rump Length	Birth-weight	Gesta-tion
FD	F	Surgical	72	184	440	144
II	F	Surgical	62	172	330	142
IV	F	Surgical	67	177	300	141
Y2	M	Natural	--	--	440	154
X6	M	Natural	--	--	390	161

Longitudinal data

The left hand maturation score data are presented in Figure 1. The control group males and females were combined for analysis due to the small number of subjects. The relation between the maturation scores and age was curvilinear. The variable age from conception and its square explained 62% of the variance in hand maturation. The females lagged behind the males as indicated by a t-test of the mean deviations from the predicted curve (p 0.01). This effect was due to one female, indicated in the figure by a black triangle.

The individual hand maturation rates of the premature subjects are shown in Figure 2. The premature infants' mean differences from the predicted curve were greater than those of the control group by the nonparametric randomization test (p = 0.0004). Although they did not contribute to the analysis due to a lack of equivalent control group data, the early assessments of short gestation infants at days 147 and 159 are shown. Day 271 was the first assessment at which a premature infant had a score in the normal range. The other premature infant completing follow-up fell within the normal range by day 355. The remaining two infants did not have normal range maturation scores at their latest assessment, days 271 and 299.

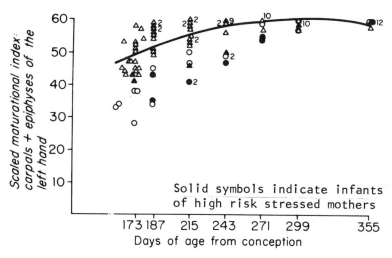

Figure 1. Scaled ossification centers in hand for normal (Δ) and short-term gestation (o) pigtail macaque infants. Solid curve = predicted values by multiple regression. Pred = -1.446 + 41 age - 0.007 age squared.

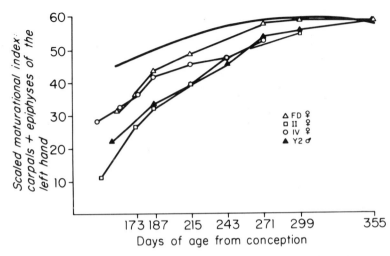

Figure 2. Scaled ossification centers in the hand for short-term gestation infants plotted against predicted curve for normals.

TABLE 3. PROPORTION OF M. nemestrina INFANTS COMPLETING HAND OSSIFICATION AND MATURATION AT THE STANDARD AGES OF ASSESSMENT

Numerator = number of infants exhibiting completed ossification or maximum maturation score; denominator = number of infants

Days from Conception	Completed Ossification		Maximum Maturation Score	
	Normal	Premature	Normal	Premature
173	1/16	0/4	0/16	0/4
187	5/15	0/4	0/15	0/4
215	9/15	0/4	4/15	0/4
343	15/16	0/4	9/16	0/4
271	9/9	2/4	7/9	0/4
299	9/9	3/3	7/9	1/3
355	9/9	2/2	9/9	2/2

Table 3 summarizes the hand data. The term-gestation group had completed ossification by day 271, with over half the infants having the full complement of 30 centers by day 215. One infant had all centers present at birth. The age range of maximum maturation score attainment is somewhat greater than the range of complete ossification, from 215 to 355 days from conception. The distribution is somewhat skewed; slightly over half the infants assessed had achieved a maturation score of 60 by day 243. On the average, the maximum maturation scores were assigned about 1 month from completed initial ossification. Age at completed initial ossification for the premature infants ranged from 271 to 299 days. Data from the two premature subjects completing follow-up suggested that maximum maturation occurs during the latter part of the normal range, from days 299 to 355. Two normal infants had anomalous carpal ossification, lacking the os centrale and trapezoid, respectively. Several other control group infants had supernumerary epiphyses of the first metacarpal.

Figure 3 illustrates the distribution of the left foot maturation scores. Again, the relation between age and the maturation scores was curvilinear, with age and age squared accounting for 50% of the variance in the maturity index. There was no sex difference in the data.

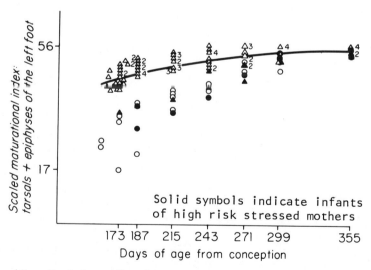

Figure 3. Scaled ossification centers in foot for normal (Δ) and short-term gestation ($\circ$) pigtail macaque infants. Solid curve = predicted values by multiple regression. Pred = 1.877 + 0.21 - 0.0003 age squared.

Figure 4 presents the individual foot maturation scores for the premature infants, which lagged behind the control group in ossification and maturation of the foot (p = 0.0001, nonparametric randomization test). In each case the mean difference scores from the predicted curve were greater for the foot than for the hand.

Table 4 summarizes the left foot ossification and maturation data. For both term and premature infants, age at complete initial ossification and maximum maturation score attainment was delayed for the foot compared with the hand. In the normal group, initial ossification was complete at ages ranging from 215 to greater than 355 days from conception. The maximum maturation score was given to the first control infant at day 243. Only half the infants assessed at day 355 had the highest maturation score of 56. None of the premature infants achieved initial ossification completion until at least day 355, and none of the premature subjects completing follow-up had a maximum maturation score.

As seen in Figure 5, the relation between foot length and age was linear, with a positive correlation of 0.87. Although the birth data showed a sex difference in foot length, multiple regression analysis indicated no difference between the sexes after controlling for age from conception.

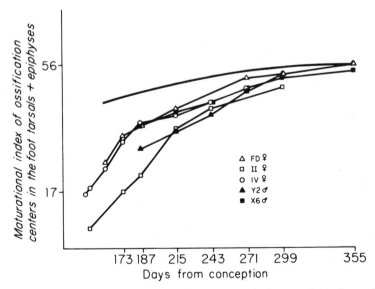

Figure 4.   Short gestation pigtail macaque infants plotted against the normal curve for skeletal maturation of the foot.

TABLE 4.   PROPORTION OF M. nemestrina INFANTS COM-
PLETING FOOT OSSIFICATION AND MATURATION
AT THE STANDARD AGES OF ASSESSMENT

Numerator = number of infants exhibiting completed ossification or
maximum maturation score; denominator = number of infants.

Days from Conception	Completed Ossification		Maximum Maturation Score	
	Normal	Premature	Normal	Premature
173	0/16	0/4	0/16	0/4
187	0/15	0/5	0/15	0/4
215	1/15	0/5	0/15	0/5
243	8/16	0/4	1/16	0/4
271	4/9	0/4	3/9	0/4
299	5/9	0/4	4/9	0/4
355	7/9	2/3	4/9	0/3

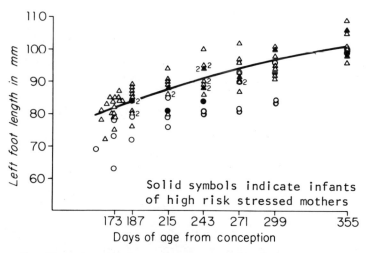

Figure 5.   Foot length for normal gestation (Δ) and short gestation (o) pigtail macaque infants plotted against predicted normal line. Pred = 4.95 + 0.23 age.

Figure 6 shows the premature infants' individual foot length growth rates. The mean difference scores of the premature subjects were again lower than the control group scores (p = 0.003, nonparametric randomization test). One infant, represented by the open triangle, was always within the normal range.

Figure 7 indicates that the relation between crown-rump length and age from conception was curvilinear, with a positive correlation of 0.96. There was a sex difference in the normal gestation group: the overall mean difference from prediction for the males was +4.04 (SE 2.35) and for all females, −5.17 (SE 2.15). Sex-specific comparisons indicated no overall differences between the premature and term-gestation infants.

DISCUSSION

These data provide a preliminary opportunity to describe physical development in infant macaques and to examine later growth and skeletal maturation in terms of gestation length and developmental status at birth. Because of small sample size, the data were analyzed by gestation length and sex. The lack of sex differences in the control group birth maturation scores and foot longitudinal data is probably a function of sample size. In a larger term-gestation sample of 24 males and 21 females, females were

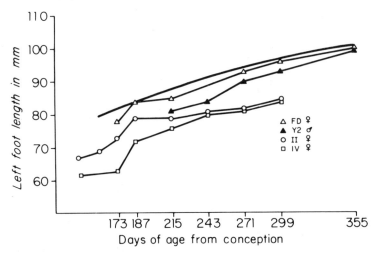

Figure 6.   Short-gestation subjects plotted against predicted normal foot length.

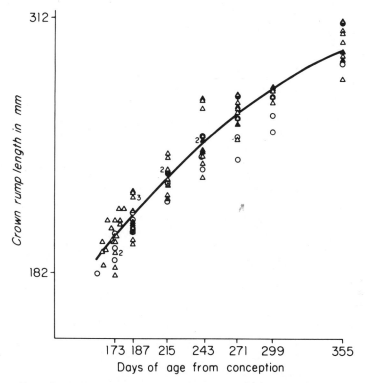

Figure 7.　Crown-rump length for normal (Δ) and short-gestation (o) pigtail macaque infants plotted against predicted curve for normals.　Pred = 2.82 + 1.24 age - 0.001 age squared.

significantly ahead of males at birth in both the number of centers ossified and the maturation score (Fahrenbruch et al., in preparation).　Unfortunately, this larger sample of animals was not available for longitudinal assessment.　A larger sample size might pick up maturational differences during infancy as well.　Although 18 of the subjects were of varied maternal risk and stress classifications, the number of infants is insufficient to examine growth and maturation by risk and stress background.　High-risk infants whose mothers were stressed by capture and handling during pregnancy are indicated in the figures, as they are probably the group most likely to have delayed ossification onset.

Skeletal growth and maturation differences may exist between spontaneously and surgically delivered premature infants. However, no clear differences existed between the five premature subjects on the basis of sex or delivery type. Short gestation does have an effect on later growth and development, as the premature infants clustered below the control group on the ossification counts and skeletal maturation scores at the standard assessment ages. The anthropometric measures were less sensitive to the effects of short gestation, and they also exhibited greater variability than the ossification measures. Although one premature infant was always within the normal range, as a group the premature infants lagged behind the control infants for foot length. The three premature females, having equivalent gestations, maintained a consistent order of catch-up to the normal range on the maturation and growth measures. The female that was heavy for gestation and had the highest maturation score at birth was the first to catch up on the ossification measures. She was always within the normal foot and crown-rump length ranges. The other two females, having lower maturity scores and normal gestation-specific birth weight, lagged farther behind the normal group on the maturation measures. They appeared to level off in a lower growth channel on the anthropometric measures. These findings may suggest the predictive value of size and developmental status at birth for the later growth and development of the premature macaque.

The utility of skeletal maturation assessment for primate nursery management is in the increased ability to identify infants with delayed maturation and potential growth problems. Weight alone is not in all cases a good index of maturity, especially if the gestation is unknown. A small infant may be very mature for size, or markedly immature skeletally. Similarly, with normal gestation and birth weights infants may, upon examination, be very immature. Radiologic assessment of the hand and foot can identify immature infants that may be at higher risk during development.

SUMMARY

A sample of 21 M. nemestrina newborns of known gestational age was assessed for physical growth and skeletal maturation status throughout the first 6 months of life. Preliminary findings indicated that short-gestation infants (less than 162 days) clustered below term-gestation infants when assessed for foot length and hand and foot ossification status at standard ages from conception. A catch-up phenomenon was observed as some premature infants fell within the normal maturation score and foot length range during the latter part of the assessment period. Gestation and sex-specific birth weight and developmental status at birth as measured by skeletal maturation may have predictive value for later growth and development.

ACKNOWLEDGMENTS

We wish to thank Drs. Laura Newell-Morris and Lewis Tarrant for their methodological suggestions, and Betsy Mitchell for her help in radiographing and measuring the infants. This work was supported by NICHD grants HD08633 and HD02274 to the Child Development and Mental Retardation Center; NIH grant RR00166 to the Regional Primate Research Center, and NIH grants SCOR-61-1417 and SCOR-61-1763 to the Department of Pediatrics at the University of Washington.

REFERENCES

Flory, C. D. Osseous development in the hand as an index of skeletal development. Monogr. Soc. Res. Child Develop. 1:1-141, 1936.

Garn, S. M. and Rohmann, C. G. Communalities of the ossification centers of the hand and wrist. Amer. J. phys. Anthrop. 17:319-323, 1959.

Garn, S. M. and Rohmann, C. G. Communalities in the ossification timing of the growing foot. Amer. J. phys. Anthrop. 24:45-50, 1968.

Kerr, G. Skeletal growth in the fetal macaque. Pp. 289-298 in: Fetal and Postnatal Cellular Growth, Hormones and Nutrition, D. B. Cheek (Ed.), New York: John Wiley and Sons, 1975.

Newell-Morris, L. and Tarrant, L. H. Ossification in the hand and foot of the macaque (Macaca nemestrina). Amer. J. phys. Anthrop. 48:441-453, 1977.

Pryor, J. W. Ossification of the epiphyses of the hand. Bull. St. College Ky. Series 3(4):1-33, 1906.

Tanner, J. M., Whitehouse, R. H., Healy, M. J. R., and Goldstein, H. Standards for Skeletal Age, Paris: Centre International de l'Enfance, 1972.

van Wagenen, G. and Asling, C. W. Ossification in the fetal monkey (Macaca mulatta): Estimation of age and progress of gestation by roentgenography. Amer. J. Anat. 114:107-132, 1964.

CHAPTER 8

AGE DETERMINATION IN MACAQUE FETUSES AND NEONATES

L. L. Newell-Morris

Departments of Anthropology and Orthodontics
University of Washington
Seattle, Washington

INTRODUCTION

Pediatricians have repeatedly stressed the need for accurate assessment of gestational age in the human fetus and newborn to ascertain neonatal risk, to permit age-controlled studies of neonatal physiology, and to provide baseline data for the developmental testing of preterm infants (Usher, McLean and Scott, 1966; Finnstrom, 1972).

Various methods for age assessment are currently used in obstetric and pediatric practice. During pregnancy, gestational age can be assessed by ultrasound estimates of fetal size, viz., from crown-rump length and biparietal diameter, or by tests on amniotic fluid. At birth, age assessment is by physical and neurological examination (Lubchenco, 1976) with birth weight and crown-rump or crown-heel length the somatometric measures most commonly used (Wylie and Amidon, 1951; Battaglia and Lubchenco, 1967; Baynall, Jones, and Harris, 1975). However, weight is an unreliable index of maturity and therefore of gestational age, primarily due to its extreme variability at any given age, especially in the latter half of pregnancy (Trolle, 1948; Finnstrom, 1972). And although linear body dimensions show better correlations with gestational age, the widespread use of crown-rump length has also been questioned as an accurate measure of fetal maturity (Farr, Kerridge, and Mitchell, 1966).

Recognizing the unreliability of age determination using somatometric measurements alone, investigators have turned to developmental criteria in their search for other variables more closely associated with gestational age. Various scoring systems

have been devised which incorporate external characteristics such
as plantar creases, skin and hair texture, earform, and testicular
descent (Usher et al., 1966; Farr et al., 1966; Farr, Mitchell,
Neligan, and Parkin, 1966); fetal and neonatal ossification status
(Christie, 1949; Russell, 1969); and neurological signs (St. Anne-
Dargassies, 1962; Amiel-Tison, 1968). Gestational age may then be
estimated from physical or neurological characteristics alone or
combined (Dubowitz, Dubowitz, and Goldberg, 1970).

Although the birth of a macaque does not involve the range
of medical and ethical issues that surround the birth of a human
infant, several considerations make the accurate assessment of
gestational age equally important in both species. For example, in
the management of the newborn monkey, maturational status is
one criterion for deciding whether the neonate will do better with
its mother or in a nursery. But how do we determine maturational
status? Weight is the most commonly used physical parameter for
classifying animals as "premature" or of "low birth weight"
(Sackett, Holm, Davis, and Fahrenbruch, 1974), but surprisingly it
has not been determined whether weight alone is the best measure
of maturity and, by extension, gestational age. Perhaps even more
importantly, the need for methods to enable accurate determination of gestational age becomes critical, as the macaque is increasingly used as a model of the human fetus in physiological testing ex
utero (Palmer, Morgan, Prueitt, Murphy, and Hodson, 1977), in
experiments in utero (Kerr, Tyson, Allen, Wallace, and Scheffler,
1972a; Hill, 1974; Riopelle, Hale, and Watts, 1976), and in research
dealing with prematurity (see Chapter 7).

Ideally, the way to obtain macaque fetuses of known gestational age is through timed matings; but timed matings, especially
when accompanied by laparoscopy, are laborious and costly to
obtain. For many studies, a fetus of an estimated age interval is
satisfactory, and monitoring of the visible estrus cycle (Blakley,
Blaine, and Morton, 1977), followed by palpation of the uterus
(Mahoney, 1975), is used to estimate time elapsed since conception.
But these assessments are quite subjective, with the possibility of
error compounded when more than one person is involved in the
evaluations. However, even when the fetus is from a timed mating
its growth and developmental status should be evaluated for
several reasons. First, errors do occur in mating schedules, in the
determination of pregnancy, and in the transfer of animal information to a computerized format as is done in large breeding colonies.
Second, it is equally important to assess the growth and developmental status of a fetus or newborn apart from chronological age.
Within any normal series of fetuses or newborns of equivalent
gestational age, there is variability of growth and development due

to genetic and environmental variables and it is important to be able to determine an animal's status relative to a normal population. When using fetuses of unknown gestational age, a developmental ranking of the fetuses may provide the most biologically meaningful series for a particular study, and finally, there should be some means for assessing the "normalcy" of a fetus of unknown age using it as its own control.

Numerous studies present basic data on macaque fetal growth and development including organ and body weights (Schultz, 1956; Dawes, Jacobson, Mott, and Shelley, 1960; Jacobson and Windle, 1960; van Wagenen and Catchpole, 1965; Fujikura and Niemann, 1967; Kerr, Kennan, Waisman, and Allen, 1969; Wilson, Fradkin, and Hardman, 1970; Mellits, Hill, and Kallman, 1975), linear dimensions (Schultz, 1937, 1956; van Wagenen and Catchpole, 1965; Kerr, Wallace, Chesney, and Waisman, 1972; Kerr, 1975; Newell-Morris, 1977), and skeletal maturation (van Wagenen and Asling, 1964; Wilson et al., 1970; Kerr et al., 1972b; Newell-Morris and Tarrant, 1978). Somatometric and skeletal maturational data are also available on the newborn macaque (Sirianni, Swindler, and Tarrant 1975; Tarrant, 1977). However, few studies have statistically analyzed the relation of growth and developmental variables with gestational age throughout the fetal period. The investigations designed to estimate fetal age from growth or developmental parameters have primarily focused on fetuses in utero using sonar biparietal diameters (Sabbagha, Turner, and Chez, 1975), radiographic skull diameters (Ferron, Miller, and McNulty, 1976), humeral and femoral length (Hutchinson, 1966), or a combination of linear measurements and skeletal maturation (Dobbelaar and Arts, 1972; Mahoney, 1975). With the exception of analyses by Mellits (1975) on rhesus macaques (Macaca mulatta) and Newell-Morris et al. (in press) on pigtail macaques (M. nemestrina), there are no studies designed to provide a quantitative method for estimating gestational age ex utero.

In brief, then, the two basic questions that confront investigators working with fetal or neonatal macaques are: 1) If an animal is of unknown conception date, what is an easy and reliable method for estimating its gestational age? 2) If an animal is of known or estimated gestational age, how does it compare to a normal population standard? With these considerations in mind, the present chapter 1) evaluates the reliability of several somatometric measures and ossification status for the estimation of gestational age in the fetal and neonatal pigtail macaque, and 2) describes the use of growth and developmental indices to determine the "normalcy" of animals of both known and unknown gestational age.

MATERIALS AND METHODS

The sample of 53 pigtail macaques included 37 fetuses (16 males, 21 females) and 16 newborns (7 males, 9 females) between 130 and 185 days gestation. All animals were of known gestational age which was calculated from the day of conception (day 1) as determined by laparoscopy; error of age assignment is considered to be no more than ± 24 hr. All fetuses were obtained by hysterotomy and all newborns were from normal vaginal deliveries.

The growth and developmental variables examined in relation to gestational age included three somatometric measurements and two skeletal maturational indices which are highly correlated with

Figure 1.   Initial positioning of a live neonate on the foot board. The most distal part of the heel is placed directly against the upright board and the foot pressed against the flat metric surface with the third toe extended.

gestational age (Newell-Morris and Tarrant, 1978; Newell-Morris, Orsini and Seed, in press). The somatometric measurements included crown-rump length, foot length, and weight; the maturational indices were derived from a scoring system applied to the number of ossified centers in the hand and foot. Unlike the human neonate, the newborn macaque exhibits most of the carpals, tarsals and secondary epiphyses of the hand and foot, ossification of these centers having begun at about 120 gestational days (Newell-Morris and Tarrant, 1978).

Crown-rump length was taken with the animal in the extended position, i.e., with the animal's back pressed against a flat surface and its head in the Frankfort plane. The greatest distance between the ischial tuberosities and the vertex of the skull was measured by sliding calipers. Foot length was measured on the left foot from the most posterior point of the heel to the distal end of the third toe using a foot board with a millimeter rule (Figure 1).

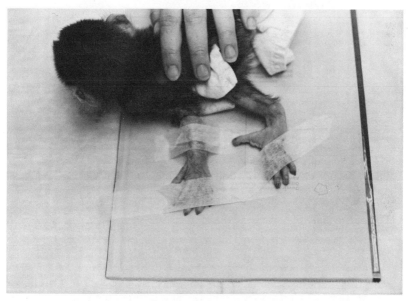

Figure 2.   Positioning of the left hand and foot of a live neonate for a posterior-anterior radiograph. Taping of the extremities insures direct contact of the volar and plantar surfaces on the radiographic film and extension of the digits; it also prevents a live animal from moving. The tape does not reduce the clarity of the x-ray.

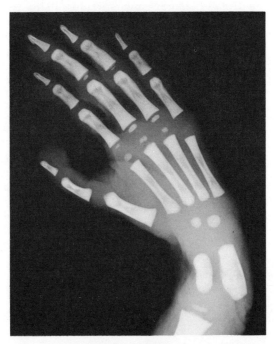

Figure 3.   Posterior-anterior radiograph of the left foot of a
            female fetus of 156 gestational days. In this case the
            NOFC score is 32.5 (base diaphyseal number 21 plus
            11.5). The arrow indicates a distal epiphyseal center of
            metacarpal 5 which is in the state of initial ossification
            (score = 0.5).

Linear measurements were taken to the nearest 0.5 mm; weight of
the animal was recorded to the nearest 0.5 g. Ossification status
was assessed from posterior-anterior radiographs taken with the
left hand and foot placed directly on nonscreen radiographic film
(Figure 2). Each carpal and tarsal bone and secondary epiphyseal
center of the long bones (phalanges, metacarpals, metatarsals and
distal radius, ulna, tibia and fibula) was assigned a score of either
0.5 (initial ossification, i.e., a barely visible, small bony nucleus),
or 1.0 (beyond the initial stage) (Figure 3). The total score for
each extremity was then obtained by summation; the score for the
hand was added to the base number of 19 primary diaphyses to
obtain the total index NOHC: the score for the foot was added to
the base number of 21 (19 diaphyses plus the calcaneus and talus

which appear between 70 and 90 gestational days) to give the index NOFC.

Regression curves were calculated using the computer polynomial function to find the curve (line) of best fit by the least squares method. The equation is $y = a + bx + cx^2 + dx^3$. When the best fit is a straight line, c, d, etc., equal 0. Regression analysis was done using age both as the independent variable (x) to construct the best-fit growth curves, and as the dependent variable (y) to yield equations for the estimation of age (Tables 1 and 2). Regression analysis was also performed using logarithmic units

TABLE 1.　REGRESSION EQUATIONS OF MATURATIONAL STATUS (NOFC, NOHC) AND SOMATOMETRIC MEASUREMENTS (WEIGHT, FOOT LENGTH, CROWN-RUMP LENGTH) ON GESTATIONAL AGE

Variable	No.	Regression Equation	S.E. of Estimate	Correlation (r)*
NOFC				
F	30	−42.8696 + 0.492066x	2.99	0.933
M	23	−56.5490 + 572163x	1.76	0.983
NOHC				
F	30	−18.3496 + 0.356148x	3.34	0.859
M	23	−52.4160 + 0.557922x	2.61	0.962
Weight (g)				
F	29	−199.023 + 4.00094x	53.93	0.766
M	20	−631.037 + 7.10353x	43.99	0.937
FL (mm)				
Combined	53	−0.0137483 + 0.459749x	4.24	0.865
C-RL (mm)				
Combined	45	59.8945 + 822167x	8.75	0.831

*Significant at $p < 0.001$

FL = foot length, C-RL = crown-rump length

leading to the regression lines log y = a + bx and y = a + b log x. The best fit lines describing each variable as a function of gestational age were tested for sex difference using analysis of covariance and an F test of significance. A correlation matrix of growth and developmental variables was calculated using 17 males and 24 females for whom all five variables were available (Table 3).

The five variables were entered into a multiple regression analysis format in which the computer determines the best estimators of age in descending hierarchical order. To evaluate attainment of a particular state of development relative to a normal standard population, I used the growth or developmental index of Drash,

TABLE 2.   REGRESSION   EQUATIONS   OF   GESTATIONAL AGE   ON   MATURATIONAL   STATUS   (NOFC, NOHC)   AND   SOMATOMETRIC   MEASUREMENTS (WEIGHT, FOOT LENGTH, CROWN-RUMP LENGTH)

Age (days) from:	Estimated No.	Regression Equation	S.E of Estimate	Corre- lation (r)*
NOFC				
F	30	95.1585 + 1.76763x	5.67	0.933
M	23	100.6050 + 1.68873x	3.02	0.983
NOHC				
F	30	76.9969 + 2.07227x	8.05	0.859
M	23	98.2466 + 1.65805x	4.49	0.962
Weight (g)				
F	29	91.1844 + 0.145624x	10.29	0.763
M	20	96.5145 + 0.123658x	5.80	0.937
FL (mm)				
Combined	53	37.9046 + 1.62822x	7.98	0.865
C-RL (mm)				
Combined	45	-4.54467 + 0.840307x	8.85	0.831

*Significant at p < 0.001

FL = foot length, C-RL = crown-rump length

Heese and Brasel (1968). The index was obtained by first calculating the estimated age of each animal in the normal sample from
each variable, using the regression equations of Table 2 to obtain
five separate estimated ages which are considered representative
of the biological age of an animal. For example, the estimated
biological age obtained using weight as the predictive variable is
the weight age. This value expresses an animal's weight as the age
at which that specific weight is statistically predicted within the
normal population. Each biological age thus obtained was then
divided by the animal's actual gestational age to yield five separate
growth and developmental indices. A perfect fit of the estimated
biological age with the actual gestational age would yield an index
of 1.00. The mean growth and developmental indices and their
standard deviations were calculated for the entire sample using
each of the five variables (Table 4).

The usefulness of the developmental indices for assessing
normalcy was tested within two groups of abnormal animals of
known gestational age. The first group consisted of eight fetuses
whose mothers had received daily dosages of retinoic acid (the
alcohol-soluble acid form of vitamin A) from days 20 to 44 of
pregnancy. The teratogenic effects of this substance have been

TABLE 3.   CORRELATION MATRIX OF GROWTH AND
            DEVELOPMENTAL VARIABLES

	Correlation Coefficient (r)*							
	Male (N=17)**				Female (N=24)**			
Variable	NOFC	NOHC	FL	Weight	NOFC	NOHC	FL	Weight
NOHC	0.96				0.94			
FL	0.96	0.96			0.86	0.79		
Weight	0.97	0.93	0.95		0.84	0.80	0.89	
C-RL	0.86	0.83	0.89	0.90	0.83	0.71	0.88	0.87

*Significant at p < 0.0001.

**Only animals were used on whom values for all five variables
    were available.

FL = foot length; C-RL = crown-rump length

TABLE 4.   MEAN GROWTH AND DEVELOPMENTAL INDICES
            (ESTIMATED AGE/ACTUAL AGE) -- NORMAL
            SAMPLE

	Male		Sex Combined		Female	
	Mean	S.D.	Mean	S.D.	Mean	S.D.
NOFC	1.000	0.024			1.001	0.036
NOHC	1.001	0.031			1.002	0.053
Weight	1.002	0.037			1.004	0.066
Foot length			1.003	0.052		
Crown-rump length			1.003	0.059		

described elsewhere (Fantel, Shepard, Newell-Morris, and Moffet,
1977).   The second group included five animals classified as
abnormal pregnancy outcome (one abortion and four stillbirths).
No animals were included in the stillbirth group if death was
attributable to birth trauma.

RESULTS

The regression equations and relevant statistics for the
curves describing the somatometric measurements and maturational indices as a function of gestational age are given in Table 1;
the best fit curves plotted from these equations are shown in
Figures 4-6.   All correlations are linear; neither higher order
polynomials nor semilogarithmic regression analyses yielded a
better fit.   As there was no statistically significant sex difference
in crown-rump or foot length, the results are presented with the
sexes combined.   The curves for weight and the two maturational
indices showed significant sex differences in both slope and elevation, and are therefore presented sex-specific.   Males were heavier
than females from approximately 150 days through term; females
were more advanced than males in skeletal maturation, with the
difference especially pronounced in the ossification status of the
hand from 130 to 160 gestational days (Figs. 5 and 6).

For both sexes the maturational index NOFC gave the highest
correlation with gestational age (females: r = 0.933, S.E. = 2.00;

males: r = 0.983, SE. = 1.76). NOHC yielded slightly lower correlation coefficients (females: r = 0.859, S.E. = 3.34; males: r = 0.96, S.E. = 2.61). Although the correlation of weight with gestational age was also high in the males (r = 0.937), it was the lowest of all correlations in the females (r = 0.766). Of the two linear measurements, foot length was slightly better associated with age (r = 0.865) than was crown-rump length (r = 0.831).

The equations derived from linear regression analysis with age as the dependent variable (y) together with the standard errors of the estimates and correlation coefficients are presented in Table 2. The maturational index NOFC gave a standard error of 5.67 days for females and 3.02 days for males; the next best

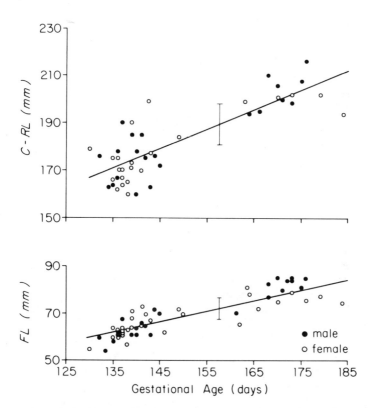

Figure 4. Relation between crown-rump length (C-RL) and foot length (FL) and gestational age. Sex-combined regression lines with standard error of the estimates from Table 2.

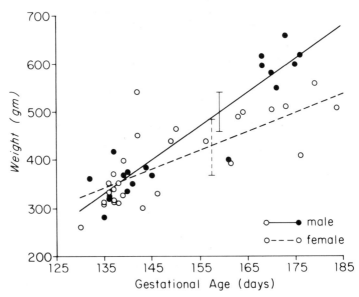

Figure 5.   Relation between weight and gestational age.   Sex-
specific regression lines with standard error of the
estimate from Table 2.

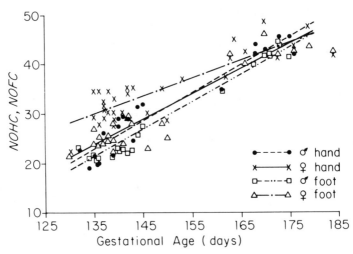

Figure 6.   Relation between the two maturational indices (NOHC,
NOFC) and gestational age.   Sex-specific regression
lines from Table 2.

predictor was NOHC (S.E. = 8.05 days for females, 4.49 days for males). Weight yielded a fairly good estimate of age for males (S.E. = 5.80 days), but not for females (S.E. = 10.29 days). The standard errors of the estimates derived from foot length and crown-rump length were 7.98 days and 8.85 days, respectively.

All variables were highly intercorrelated (r = 0.71-0.97), significant at p < 0.0001 (Table 3). Males showed consistently higher correlation coefficients (r = 0.83-0.97) than did females (r = 0.71-0.94). However, the rank ordering of the correlation coefficient values across all variables was the same for both sexes.

Entry of all five variables into a multiple regression analysis format did not significantly improve the standard error of the age estimate obtained with NOFC alone. Thus, for both sexes the ossification status of the foot is taken here as the most reliable variable for estimating the age of a fetal or newborn macaque.

In choosing which variable to use when estimating age, a low standard error is one criterion, but the variable chosen should also be minimally affected by environmental factors. Such a consideration is crucial because if an environmental insult occurs, any age estimate obtained using an affected variable may be significantly less than the actual gestational age. To determine which of the five variables used here were the most resistant to fetal stress, the two groups of abnormal animals (retinoic acid animals and abnormal pregnancy outcome) were taken as representative of animals who had experienced environmental stress. Their five growth and developmental indices were then compared with the mean indices from the normal population. Ossification status appeared to be most resistant to fetal stress. Six of the eight retinoic acid animals exhibited both weight and foot length indices 2 S.E. or more below the normal population indices; of these, four animals also had a crown-rump length index below 2 S.E. (Table 5). Unfortunately, only three of the eight animals could be assessed for ossification status; the other five animals had some degree of ectrodactyly (absence of all or part of a digit). Of the three animals with normal limbs, none had a foot developmental index below 2 S.D. and only one had a hand index below 2 S.E. In the male of 162 gestational days, growth had been significantly depressed (indices less than 2 S.D.) whereas skeletal maturation seemed unaffected. The female of 161 gestational days, although less retarded in both growth and development, fell below 1 S.D. in the growth indices, but not in the developmental indices. The male of 165 days is difficult to assess; foot length was not available, the hand index was below 2 S.D., weight and NOFC indices were within 1 S.D., and crown-rump length was above the normal population mean.

The animals in the abnormal pregnancy outcome group presented some interesting contrasts. The single abortion, a 132-day male, seemed normal in all respects, in fact slightly above the normal population mean index for NOFC, weight and crown-rump length. Of the four stillbirths, one (a 180-day male) was retarded in growth and development, whereas a female of 180 days was normal in all parameters. The indices of a 170-day female were slightly depressed (below 1 S.D.), but not enough to classify the animal as abnormal. The 167-day female presents an interesting case in that the maturational indices were well within normal

TABLE 5.   DEVELOPMENTAL INDICES (ESTIMATED AGE/AGE)
            -- ABNORMAL SAMPLES

Age	Sex	NOFC*	NOHC*	Weight	Foot Length	Crown-rump Length	Comments
RETINOIC FETUSES (N = 8)							
132	M	--	--	0.926^2	0.867^2	1.080	limb deformities
144	M	--	--	0.881^3	0.862^2	0.972	abortion limb deformities
162	M	0.975^1	0.985	0.825^3	0.847^3	0.854^2	
163	M	--	--	0.687^3	0.712^3	0.787^3	abortion
165	M	0.968^1	0.937^2	0.933^1	--	1.062	stillbirth
152	F	--	--	0.828^2	0.699^3	0.772^3	limb deformities
161	F	0.981	0.954	0.889^1	0.923^1	0.937^1	
185	F	--	--	0.868^2	0.839^3	0.802^3	limb deformities
ABNORMAL PREGNANCY OUTCOME (N = 5)							
132	M	1.031	0.995	1.021	0.935^1	1.003	abortion
180	M	0.920^2	0.951^1	0.832^3	0.871^2	0.894^1	stillbirth
167	F	1.014	0.970	0.909^1	0.885^2	0.899^1	stillbirth (mother ill)
170	F	0.960^1	0.983	0.937^1	0.932^1	--	stillbirth
180	F	0.988	1.003	1.041	0.989	--	stillbirth

*Maturational indices were not calculated on animals with limb deformities which included ectrodactyly.

[1] $\leq$1 S.D.
[2] $\leq$2 S.D.
[3] $\leq$3 S.D.

means whereas the growth indices fell below 1 S.D. This animal's mother had been ill for at least two weeks before parturition, and blood had been drawn on several occasions during the last third of her pregnancy.

Calculation of the developmental indices depends on knowing the actual gestational age, leaving unanswered the question of how to determine the normalcy of an animal of unknown gestational age. Recognizing that there is large room for error, I propose the following approach which assumes that the parameters of growth and development are well integrated in a normal animal. Thus, the biological ages calculated from the equations for NOHC, weight, foot length and crown-rump length should approximate the biological age estimate given by the best age estimator, i.e., NOFC. The estimated gestational age obtained using NOFC is taken as the biological age most representative of the animal's actual gestational age. The three growth biological ages are calculated using weight, foot length and crown-rump length and then averaged to obtain a mean growth biological age. If the latter age falls below the 2 S.E. of the biological age estimate obtained from NOFC, the

TABLE 6.   RETINOIC   ACID   ANIMALS   COMPARED   WITH MATCHED NORMALS FOR BIOLOGICAL INTEGRATION OF GROWTH AND DEVELOPMENT SYSTEMS

Age (days)	Sex	Preg. Class	Estimated Biological Age				Foot B.A. -2 S.E.*	Mean Growth B.A.
			NOFC	Wt	FL	C-RL		
162	M	R	158	134	137	138	(152)	136**
168	M	N	172	173	173	172	(166)	173
165	M	R	160	155	---	175	(154)	165
168	M	N	172	173	172	173	(166)	173
161	F	R	158	143	149	151	(147)	147**
163	F	N	168	162	168	162	(151)	164

Wt = weight, FL = foot length, C-RL = crown-rump length, B.A. = biological age, R = retinoic acid, N = normal

*S.E. of estimate for age for NOFC = 3.02 (male) and 5.67 (female)

**At or below 2 S.E. of the foot ossification biological age normalcy of the animal is suspect. Applying this method to the retinoic acid animals, it is possible to detect two animals in which the developmental indices differ considerably in value from the growth indices, and would therefore be classified as abnormal. However, the third animal would be accepted as normal. None of the normal animals selected for controls showed this disintegration of growth and development, as the mean growth biological age was consistently well within the 2 S.E. of the foot biological age (Table 6).

DISCUSSION

The results of this analysis show that of the five growth and developmental variables assessed, ossification status, particularly of the foot, is the best predictor of gestational age in the fetal and neonatal macaque (130 days - birth), yielding a S.E. of age estimation of less than a week in both males and females. This value is well within the error of prediction of 1.02 weeks obtained in the human fetus and newborn using a scoring system incorporating several neurological and physical variables (Dubowitz et al., 1970). The more commonly used somatometric measurements gave significantly higher standard errors of the estimate. Weight is an especially poor age predictor in females in which the S.E. is 10.29 days compared with the S.E. of 5.67 days for foot ossification. The fact that ossification status in the hand and foot is highly correlated with gestational age is not surprising in view of the proven utility of skeletal maturation for the assessment of biological development in the human child (Tanner, 1962; Mellits, Dorst and Cheek, 1971). However, due to the skeletal immaturity and therefore the lack of secondary centers in the human fetus compared with that of the macaque, assessment of gestational age in the fetal or neonatal human from osseous development is not sufficiently precise to be of value (Falkner, 1971).

Mellits (1975) obtained good gestational age estimates in the macaque (S.D. = 4.0 days) from a multiple regression equation which included five variables ranging from total body weight to DNA concentration in the cerebrum. However, the measurements used required complex assays and destruction of the fetus. The method given here of scoring hand and foot radiographs is a simple one, and requires no knowledge of the specific bones involved nor any use of subtle morphological criteria. The only judgment to be made is that of differentiating between a bone in an initial state of ossification (score = 0.5) and one beyond that stage (score = 1.0). The distinction is easily made, and in my laboratory interobserver correlations are very high.

Importantly, bone maturation also appears to be more resistant to environmental stress than are the growth parameters of length and weight. This finding agrees with the results of malnutrition studies in human infants and children which show that a hierarchy of responses by the various body tissues occurs in the presence of nutritional stress. First depleted are the fat deposits, followed by muscle mass and deficits in linear bone growth; last affected are skeletal maturation and brain growth and development (Acheson, 1960; McFie and Welbourn, 1962). Experimentally growth-retarded macaque fetuses show a similar, although not identical, hierarchy with skeletal mass less affected than muscle mass (Hill, 1974) and the brain minimally so (Hill, Myers, Holt, Scott, and Cheek, 1971), but interestingly, weight in the fetal macaque seems to be neither more nor less depressed than is length.

The lack of weight depression in the stressed macaque fetus is contrary to the findings in cases of human intrauterine growth retardation in which body weight shows an earlier and more severe response to nutritional stresses than does body length (Naeye and Kelley, 1966; Hill et al., 1971). These species differences in response probably reflect the differences between the monkey and human fetus in the amount of subcutaneous fat present. Whereas the normal human fetus rapidly acquires fat near term, the normal newborn macaque has very little, so that its weight reflects primarily bone and muscle. Therefore, when weight reduction in the monkey occurs it may reflect more acute long-term deprivation in utero than does weight reduction in the human fetus, and should alert an investigator to a highly suspect condition. Although fetal weight in a macaque may thus be less affected by environmental stress than is human fetal weight, the analysis here indicates that a more accurate assessment of maturity is obtained from the ossification status of the animal. Given the wide range of weight at any age, especially in females, as observed in our normal sample and in other studies (Fujikura and Niemann, 1967), its reliability for determining maturational status and gestational age is suspect, especially in the late fetal period. Thus, weight probably should not be used alone in making judgments regarding maturity of newborns.

Although the somatometric measurements are not as reliable predictors of gestational age as are other variables in this study, they may provide valuable information when used in conjunction with the maturational index of the foot. As has been demonstrated here, the use of developmental and growth indices derived from the biological age estimates permits comparison between abnormal

animals of known age and a normal population standard. Differences in the biological age indices obtained from the five growth
and developmental variables may give valuable clues regarding the
timing and duration of intrauterine stress. Although the evidence
is scanty, it seems that retinoic acid administered early in the
embryonic period produced multiple skeletal abnormalities and
interfered with growth (as measured by weight and length), but had
minimal effects on the progress of skeletal maturation.

The results of comparing the biological age indices of animals
labeled as abnormal pregnancy outcome with normal animals lead
to some interesting speculations. First, the aborted fetus appeared
to have been growing and developing normally, causing one to
wonder why it was aborted. Second, the stillbirths comprised a
very diverse lot ranging from a severely growth-retarded male to a
female apparently completely normal, yet they were all well within
normal parturition time. These data suggest that so-called "high
risk" females, who have higher than average rates of stillbirths,
may comprise several different classes; fetal growth and developmental status may provide a key to categorizing these females for
studies on the etiology of stillbirth.

When dealing with fetuses of unknown gestational age, the
use of the biological age concept provides a way of not only
estimating age but also of roughly assessing whether a fetus is
normal. Assuming that in a normal animal maturational and
growth systems are well integrated, an obvious difference between
age estimates provided by the equations using NOFC, and those
provided by the growth variables, suggest abnormalcy. Finally, by
combining the biological ages obtained, or by using foot maturational age alone (if the animal seems to be normally integrated), an
investigator can rank any series of animals of unknown gestational
age into a biologically meaningful series of utility for innumerable
research projects.

In summary, once the NOFC score has been obtained, use of
the regression equation presented here yields a good biological age
estimate for a macaque fetus or newborn. This calculation in turn
allows an investigator to 1) check the estimated biological age
against the actual gestational age, when known, to detect possible
errors in any one of the numerous steps involved in obtaining a
known-aged fetus; 2) detect an abnormal animal if there is a
signficant discrepancy between the estimated biological age and
actual gestational age; 3) detect an abnormal animal when actual
gestational age is not known; and 4) assess the physiological
maturation of an animal, which may be more predictive of survival
chances than is weight, which is not the best predictor of gestational age, especially in the female.

A note of caution must be introduced regarding the data used here to derive the statistical formulations. First, the sample was small, and subject to all errors inherent to this fact. Second, there was not an even distribution of animals over all age groups. Thus, although a linear regression equation gave the best fit curve, a larger sample with more animals aged 145-165 days might yield curvilinear relationships of weight and crown-rump length as have been demonstrated in the human fetal data (Lubchenco, Hansman, Dressler, and Boyd, 1963). Lastly, care must be taken not to extrapolate beyond the age period used here (130-185 days) as the growth, and especially the developmental curves, change on either side of this age interval.

There is also a question whether the equations presented here may be applied with some confidence to rhesus macaques. Although admittedly the error of estimate would be increased slightly, I believe, based on a comparison of published rhesus macaque data with our own (Newell-Morris et al., in press; Newell-Morris and Tarrant, 1978), that they may be applied to the rhesus macaque. Obviously, comparative studies of the fetal growth and development of several species of macaque are needed, especially to yield statistical parameters that are useful in the ways I have indicated here.

REFERENCES

Acheson, R. M. Effects of nutrition and disease on human growth. Pp. 73-92 in: Human Growth, J. M. Tanner, (Ed.), Oxford: Pergamon Press, 1960.

Amiel-Tison, C. Neurological evaluation of the maturity of newborn infants. Arch. Dis. Child. 43:89-93, 1968.

Baynall, K., Jones, P., and Harris, P. Estimating the age of the human fetus from crown-rump measurements. Ann. hum. Biol. 2:387-390, 1975.

Battaglia, F. C. and Lubchenco, L. O. A practical classification of newborn infants by weight and gestational age. J. Pediat. 71:159-163, 1967.

Blakley, G. A., Blaine, C. R., and Morton, W. R. Correlation of perineal detumescence and ovulation in the pigtail Macaca nemestrina. Lab. anim. Sci. 27:352-355, 1977.

Christie, A. Prevalence and distribution of ossification centers in the newborn infant. Amer. J. Dis. Child 77:355-361, 1949.

Dawes, G. S., Jacobson, H. N., Mott, J. C., and Shelley, H. J. Some observations on foetal and newborn rhesus monkeys. J. Physiol. 152:271-298, 1960.

Dobbelaar, M. J. and Arts, T. H. M. Estimation of the foetal age in pregnant rhesus monkeys (Macaca mulatta) by radiography. Lab. anim. Sci. 6:235-240, 1972.

Drash, A., Heese, D., and Brasel, J. A. Clinical material: Anthropometric and developmental analysis. Pp. 60-83 in: Human Growth, D. B. Cheek (Ed.), Philadelphia: Lea and Febiger, 1968.

Dubowitz, L., Dubowitz, V., and Goldberg, C. Clinical assessment of gestational age in the newborn infant. J. Pediat. 77:1-10, 1970.

Falkner, F. Skeletal maturity indicators in infancy. Am. J. phys. Anthropol. 35:393-394, 1971.

Fantel, A. G., Shepard, T. H., Newell-Morris, L. L., and Moffett, B. C. Teratogenic effects of retinoic acid in pigtail monkeys (Macaca nemestrina). Teratology 15:65-72, 1977.

Farr, V., Kerridge, D. F., and Mitchell, R. G. The value of some external characteristics in the assessment of gestational age at birth. Develop. Med. Child Neurol. 8:657-660, 1966.

Farr, V., Mitchell, R. G., Neligan, G. A., and Parkin, J. M. The definition of some external characteristics used in the assessment of gestational age in the newborn infant. Develop. Med. Child Neurol. 8:507-511, 1966.

Ferron, R. R., Miller, R. S., and McNulty, W. P. Estimation of fetal age and weight from radiographic skull diameters in the rhesus monkey (Macaca mulatta). J. med. Primat. 5:41-48, 1976.

Finnstrom, O. Studies on maturity in newborn infants. VI. Comparison between different methods for maturity estimation. Acta paediat. scand. 61:33-41, 1972.

Fujikura, T. and Niemann, W. H. Birth weight, gestational age, and type of delivery in rhesus monkeys. Am. J. Obstet. Gynec. 97: 76-80, 1967.

Hill, D. C. Experimental growth retardation in rhesus monkeys. Pp 99-125 in: Size at Birth, Elsevier, Amsterdam: Ciba Foundation Symposium 27 (new series), 1974.

Hill, D. E., Myers R. E., Holt, A. B., Scott, R. E. and Cheek, D. B. Fetal growth retardation produced by experimental placental insufficiency in the rhesus monkey. II. Chemical composition of the brain, liver, muscle and carcass. Biol. Neonat. 19:68-82, 1971.

Hutchinson, T. C. A method for determining expected parturition date of rhesus monkey, (Macaca mulatta). Lab. anim. Care 16:93-95, 1966.

Jacobson, H. N. and Windle, W. F. Observations on mating, gestation, birth and postnatal development of Macaca mulatta. Biol Neonat., 3:105-120, 1960.

Kerr, G. R. Skeletal growth in the fetal macaque. Pp 289-298 in: Fetal and Postnatal Cellular Growth, D. B. Cheek (Ed.), N.Y.: John Wiley and Sons, 1975

Kerr, G. R., Kennan, A. L., Waisman, H.A., and Allen, J.R. Growth and development of the fetal rhesus monkey. I. Physical growth. Growth 33:201-213, 1969.

Kerr, G. R., Tyson, I. B., Allen, J. R., Wallace, J. H., and Scheffler, G. Deficiency of thyroid hormone and development of the fetal rhesus monkey. II. Effect on physical growth, skeletal maturation and biochemical measures of thyroid function. Biol. Neonat. 21:282-295, 1972.

Kerr, G. R., Wallace, J. H., Chesney, C. F. and Waisman, H. A. Growth and development of the fetal rhesus monkey III. Maturation and linear growth of the skull and appendicular skeleton. Growth 36:59-76, 1972.

Lubchenco, L. O. The High Risk Infant. Vol. XIV in Major Problems in Clinical Pediatrics, A.G. Schaffer and M. Markowitz (Eds.), Philadelphia: W.B. Saunders Co., 1976.

Lubchenco, L. O., Hansman, C., Dressler, M., and Boyd, E. Intrauterine growth as estimated from liveborn birth-weight data at 24 to 42 weeks of gestation. Pediatrics 32:793-800, 1963.

McFie, J. and Welbourn, H. L. Effect of manlnutrition in infancy on the development of bone, muscle and fat. J. Nutrition 76:97-105, 1962.

Mahoney, C. J. Practical aspects of determining early pregnancy. State of foetal development and imminent parturition in the monkey (Macaca fascicularis). Anim. Handbooks 6:261-274, 1975.

Mellits, E. D. Assessment of the biological age index in the fetal primate. Pp. 381-385 in: Fetal and Postnatal Cellular Growth. D. B. Cheek (Ed.), New York: John Wiley and Sons, 1975.

Mellits, E. D., Dorst, J. P., and Cheek, D. B. Bone age: Its contribution to the prediction of maturational or biological age. Am. J. phys. Anthrop. 35:381-384, 1971.

Mellits, E. D., Hill, D. E. and Kallman, C. H. Growth of visceral organs in the fetus. Pp. 209-231 in: Fetal and Postnatal Cellular Growth, D.B. Cheek (Ed.), New York: John Wiley and Sons, 1975.

Naeye, D. K. and Kelley, A. Judgment of fetal age. III. The pathologist's evaluation. Pediat. Clin. N. Am., 13:849-862, 1966.

Newell-Morris, L. Growth and developmental parameters associated with gestational age in the Macaca nemestrina. Am. J. phys. Anthrop. 47:152, 1977.

Newell-Morris, L., Arsini, J., and Seed, J. Estimation of gestational age from somatometric measurements of the fetal pigtail macaque (Macaca nemestrina) Lab. anim. Sci., in press.

Newell-Morris, L. and Tarrant, L. H. Ossification in the hand and foot of the macaque (Macaca nemestrina) I. General features. Am J. phys. Anthrop. 48:441-454, 1978.

Palmer, S., Morgan, T. E., Prueitt, J. L., Murphy, J. H. and Hodson, W. A. Lung development in the fetal primate, Macaca nemestrina. II. Pressure-volume and phospholipid changes. Pediat. Res. 11:1015-1021, 1977.

Riopelle, A. J., Hale, P. A., and Watts, E. S. Protein deprivation in primates: VII. Determinants of size and skeletal maturity at birth in rhesus monkeys. Hum. Biol. 48:203-222, 1976.

Russell, J. G. B. Radiological assessment of fetal maturity. J. Obstet. Gynaec. Brit. Cwlth 76:208-219, 1969.

Sabbagha, R. E., Turner, J. H. and Chez, R. A. Sonar biparietal diameter growth standards in the rhesus monkey. Am. J. Obstet. Gynec. 121:371-374, 1975.

Sackett, G. P., Holm, R. A., Davis, A. E., and Fahrenbruch, C. E. Prematurity and low birth weight in pigtail macaques: Incidence, prediction and effects on infant development. Pp. 189-205 in: Symposium of the 5th Congress of the International Primatological Society, Tokyo: Japan Science Press, 1974.

St. Anne-Dargassies, S. The fullterm newborn: Neurologic assessment. Biol. Neonat. 4:174-179, 1962.

Schultz, A. Fetal growth and development of the rhesus monkey. Contrib. Embryol. 26:73-97, 1937.

Schultz, A. H. 1956. Postembryonic age changes. Pp. 887-964 in: Primatologia, Vol. I, H. Hofer, A. H. Schultz, and D. Starck (Eds.), Basel: S. Karger, 1956.

Sirianni, J. E., Swindler, D. R., and Tarrant, L. H. Somatometry of newborn Macaca nemestrina. Folia Primat. 24:16-23, 1975.

Tanner, J. M. Growth at Adolescence, 2nd ed. London: Blackwell Scientific Publications, 1962.

Tarrant, L. Sex differences in skeletal maturation at birth in the pig-tailed monkey. Am.J. phys. Anthrop. 47:163, 1977.

Trolle, D. Age of foetus determined from its measures. Acta obstet. gynec. scand. 27:327-337, 1948.

Usher, R., McLean, F., and Scott, K. E. Judgment of fetal age. II. Clinical significance of gestational age and an objective method for its assessment. Pediat. Clin. Am. 13:835-848, 1966.

van Wagenen, G. and Catchpole, H. R. Growth of the fetus and placenta of the monkey (Macaca mulatta). Am. J. phys. Anthrop. 23:23-34, 1965.

van Wagenen, G. and Asling, C. W. Ossification in the fetal monkey (Macaca mulatta). Estimation of age and progress of gestation by roentgenography. Am. J. Anat. 114:107-132, 1964.

Wilson, J. G., Fradkin, R., and Hardman, A. Breeding and pregnancy in rhesus monkeys used for teratological testing. Teratology 3:59-72, 1970.

Wylie, B. and Amidon, B. F. Correlation of weight, length, and time-factors in fetal age. Am. J. Obstet. Gynec. 61:193-196, 1951.

CHAPTER 9

SERUM BILIRUBIN LEVELS IN FULL-TERM AND PREMATURE Macaca nemestrina

T. M. Burbacher and G. P. Sackett

Department of Psychology,
Regional Primate Research Center, and
Child Development and Mental Retardation Center
University of Washington
Seattle, Washington

INTRODUCTION

Neonatal jaundice is one of the most common problems among human neonates. It is estimated that 25-50% of all normal newborn infants exhibit clinical jaundice during the first week of life (Maisels, 1975). The large majority of these infants demonstrate what is called "physiological jaundice" with bilirubin concentrations reaching 5-7 mg/100 ml. Others, including the majority of premature infants, show a prolonged elevation in bilirubin concentrations (hyperbilirubinemia) with serum bilirubin levels reaching well over 20 mg/100 ml. The association between prolonged elevations in serum bilirubin levels and the occurrence of kernicterus is well documented (Hsia, Allen, Gellis, and Diamond, 1952; Perlstein, 1960). However, it is still unclear what level of bilirubin constitutes a potential risk for developing neurological or behavioral abnormalities owing to bilirubin intoxication. Several studies have found a positive relation between small increases in bilirubin concentrations and subsequent neurological and behavioral impairments (O'Dell, Storey, and Rosenberg, 1970; Hardy and Peeples, 1971). According to one study, this relation becomes critical at a concentration of 16-19 mg/100 ml (Boggs, Hardy, and Fraiser, 1967). That the potential for behavioral deficits due to bilirubin intoxication increases on a continuum with increases in bilirubin concentrations must be further considered and investigated. Factors relating to increased susceptibility to neurological or

117

behavioral abnormalities at relatively low concentrations of bilirubin (prematurity, Hyaline Membrane Disease) must also be elucidated.

The occurrence of physiological jaundice in the newborn rhesus macaque (Macaca mulatta) has been noted in previous investigations (Behrman and Hibbard, 1964; Gartner, Lee, Vaisman, Lane, and Zarafu, 1977). Gartner et al. (1977) clearly demonstrated a biphasic increase in bilirubin concentrations in newborn rhesus monkeys similar to one characteristic of human newborns. However, there is little information concerning the occurrence of abnormal levels of bilirubin or hyperbilirubinemia in nonhuman primate species.

For the past two years we have been measuring the concentration of bilirubin in newborn pigtail macaques (M. nemestrina). Bilirubin and albumin concentrations are obtained from full-term, premature, and/or low-birth-weight pigtail macaques to obtain normative data for this species as well as to investigate the possibility that a condition resembling hyperbilirubinemia may occur in the premature or immature macaque.

METHODS

Twenty-six newborn monkeys were studied. Eighteen full-term and two premature subjects were obtained from established research projects at the Infant Primate Research Laboratory at the University of Washington. Infants were defined as full-term or premature on the basis of known gestation ages resulting from timed-mating procedures. The mean gestation age for the full-term infants was 170 ±6 days. Birth weights for this group ranged from 420 to 600 g, normal for pigtail monkeys (colony statistics for the Primate Field Station of the Regional Primate Research Center at the University of Washington, located at Medical Lake, WA). The premature infants were beyond 1 S.D. for gestation-age norms generated from a larger sample of 50 timed matings. One premature infant (161 days) was also low-birth-weight while the other (154 days) was not. Six low-birth-weight animals were obtained from the lowest 10th centile of birth weights for monkeys born at the Medical Lake breeding facility. The low-birth-weight infants were not products of timed conceptions. Estimates of their gestational ages from radiographs and palpation ranged from 135 to 150 days (Beamer, Blakley, Fahrenbruch, and Sackett, in preparation) and all were designated as "premature." All infants were delivered spontaneously, separated from their mothers and reared in the nursery.

Blood samples for total bilirubin, direct bilirubin and albumin determinations were obtained during the first 2 weeks of life (see Figure 1). Blood taken from the umbilical cord at birth was used for zero-hour determinations. Samples thereafter were obtained

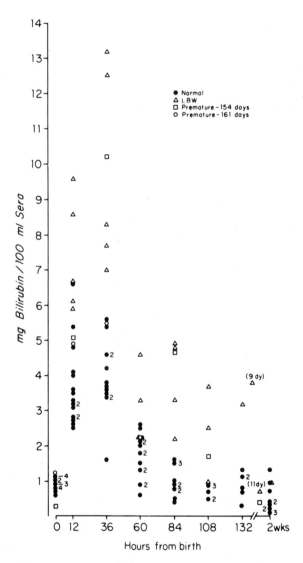

Figure 1.   Total serum bilirubin concentrations for normal, low-birth-weight, and premature pigtail macaques.

via femoral venipuncture using a 1-cc syringe with a 26-gauge needle. Bilirubin determinations were performed according to the procedure outlined by Lathe and Rothuen (1958). Albumin concentrations were obtained following the method described by Basil (1971).

RESULTS

The distribution of total serum bilirubin concentrations during the 2-week period is shown in Figure 1. Although every effort was made to collect blood from all subjects according to the designated schedule, some blood samples could not be obtained. However, no effects (e.g., higher levels of bilirubin) were correlated with repeated blood sampling. Full-term infants showed a marked increase in serum bilirubin during the first 12 hr of life. During the next 24 hr, serum bilirubin concentrations generally rose again. A rapid decline in serum bilirubin occurred for the full-term infants during the 36-hr postnatal period, followed by a slower decline and finally a leveling off from 84 to 132 hr of age. By 2 weeks of age, serum bilirubin concentrations were within the normal range for adult pigtail macaques (Morrow and Terry, 1972).

Serum bilirubin concentrations for low-birth-weight infants were generally much higher than those obtained from normal infants. A randomization test was performed on the data from the low-birth-weight and normal subjects at each age from 12 to 108 hr to calculate the exact probability of obtaining scores from the two groups ranked as displayed in Figure 1 (Siegel, 1956). Results of the randomization tests are shown in Table 1. Probabilities were not obtained for other ages owing to the small sample size of the low-birth-weight group.

TABLE 1.   RESULTS OF RANDOMIZATION TESTS ON SERUM
             BILIRUBIN CONCENTRATIONS

Hours from birth	12	36	60	84	108
probability* =	.0001	.0002	.004	.0002	.01

*Exact probability of obtaining distribution of serum bilirubin levels for low-birth-weight and normal monkeys as shown in Figure 1.

Figure 2 shows the 1 S.D. limits for the serum bilirubin concentrations from normal infants compared with the data from the low-birth-weight and premature subjects. Generally low-birth-weight infants displayed elevated bilirubin concentrations throughout the first 9 days of life. The premature infant with the gestation age of 154 days had serum bilirubin concentrations similar to those of the low-birth-weight infants while the other premature infant displayed serum bilirubin concentrations just beyond the 1 S.D. range at 12 and 36 hr only.

Mean serum albumin concentrations for full-term infants are shown in Figure 3. Albumin concentrations changed only moderately during the 2-week period and were within the normal range for adult macaques (Robinson and Ziegler, 1968). Serum albumin concentrations for low-birth-weight and premature infants were also generally within the normal range.

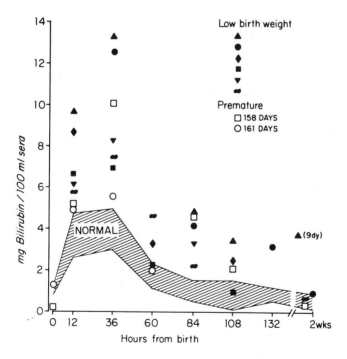

Figure 2.   Mean (±S.D.) total serum bilirubin concentrations for normal monkeys and total serum bilirubin concentrations for low-birth-weight and premature pigtail macaques.

DISCUSSION

The full-term infant monkeys examined in this study revealed a pattern of "physiological jaundice" much like that in human newborns (Gartner et al., 1977). This phenomenon was accelerated in the monkeys and the absolute levels of serum bilirubin were somewhat lower than what is seen in human infants (Gartner et al., 1977). But the typical temporary elevation of serum bilirubin levels that is commonly observed in human neonates was also seen in these newborn pigtail macaques.

The low-birth-weight infants and the premature infant of 154 gestational days displayed higher bilirubin levels, relative to normals, much like immature humans (Ackerman, Dyer, and Laydorf, 1970). The hyperbilirubinemia was accompanied by a marked discoloration of the skin in most cases. In the premature infant of 161 gestational days, serum bilirubin was elevated at first but fell within normal limits fairly rapidly. This, along with the estimates of early gestation ages for the low-birth-weight animals, suggests that the hyperbilirubinemia displayed by these monkeys is due primarily to their immaturity and is not a factor of birth weight alone.

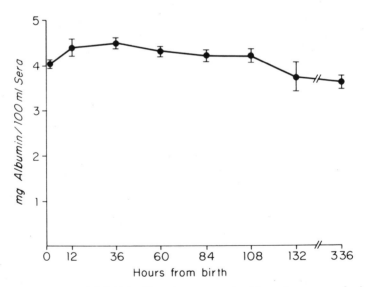

Figure 3.   Mean (± S.D.) albumin concentrations for normal pigtail macaques.

Follow-up examinations of all infants are now underway and data concerning early reflex behavior, physical growth and skeletal maturation, social behaviors and general learning capabilities will soon be available to assess the immediate as well as long-range effects of elevated bilirubin concentrations in these pigtail monkeys. Future investigations into the factors involved in this increase in bilirubin concentrations as well as possible preventative therapy may be directly applicable to human nursery care.

ACKNOWLEDGMENTS

This work was supported by NIH grants HD08633 and HD02274 from NICHD Mental Retardation Branch and RR00166 from DRR, Animal Resources Branch.

REFERENCES

Ackerman, B., Dyer, G., and Laydorf, M. Hyperbilirubinemia and kernicterus in small premature infants. Pediatrics 45:918, 1970.

Basil, C. Albumin standards and the measurement of serum bilirubin with bromocresol green. Acta clin. Chem. 31:87-96, 1971.

Beamer, T., Blakley, G., Fahrenbruch, C., and Sackett, G. Estimation of gestational age and prediction of parturition date from fetal radiographs in M. nemestrina. In preparation.

Behrman, R. and Hibbard, E. Bilirubin: Acute effects in newborn rhesus monkeys. Science 144:645, 1964.

Boggs, T., Hardy, J. and Frazier, T. Correlation of neonatal serum total bilirubin concentrations and developmental status at age eight months. J. Pediat. 71:553, 1967.

Gartner, L., Lee, K., Vaisman, S., Lane, D., and Zarafu, I. Development of bilirubin transport and metabolism in the newborn rhesus monkey. J. Pediat. 90:513, 1977.

Hardy, J., and Peeples, M. Serum bilirubin levels in newborn infants. Distributions and associations with neurological abnormalities during the first year of life. Johns Hopkins Med. J. 128:265, 1971.

Hsia, D., Allen, F., Gellis, S., and Diamond, L. Erythroblastosis fetalis studies on serum bilirubin in relation to kernicterus. New Engl. J. Med. 247:668, 1952.

Lathe, R. and Rothuen, J. Factors affecting the rate of coupling of bilirubin in the Van den Burgh reaction. J. clin. Pathol. 11:155-157, 1958.

Maisels, M.   Neonatal Jaundice.   Pp. 335-378 in: Neonatology,
    Pathophysiology and Management of the Newborn,   G. Avery
    (Ed.), Philadelphia: J. B. Lippencott, Co., 1975.
Morrow, A. and Terry, M.   Liver function tests in blood of
    nonhuman primates tabulated from the literature. A publica-
    tion of the Primate Information Center, Regional Primate
    Research Center at the University of Washington, 1972.
O'Dell, G., Storey, B., and Rosenberg, L.   Studies in kernicterus:
    III.   The saturation of serum proteins with bilirubin during
    neonatal life and its relationship to brain damage at five
    years. J. Pediat. 76:12, 1970.
Perlstein, M.   The clinical syndrome of kernicterus.   Pediatric
    Clinics of North America. Neuropediatrics 7:665, 1960.
Robinson, F. and Ziegler, R.   Clinical laboratory data derived from
    102 Macaca mulatta. Lab. anim. Care 18:50-57, 1968.
Siegel, S.   Nonparametric Statistics for the Behavioral Sciences.
    New York: McGraw-Hill Book Company, 1956.

CHAPTER 10

DEVELOPMENT OF BASIC PHYSIOLOGICAL PARAMETERS AND SLEEP-WAKEFULNESS PATTERNS IN NORMAL AND AT-RISK NEONATAL PIGTAIL MACAQUES (Macaca nemestrina)

G. P. Sackett, C. E. Fahrenbruch and
G. C. Ruppenthal

Regional Primate Research Center and
Child Development and Mental Retardation Center
University of Washington
Seattle, Washington

INTRODUCTION: AN "EDITORIAL" FORWARD

Poor pregnancy outcomes are a major problem for good primate colony husbandry, yet they are a major scientific asset for a number of nonhuman primate research goals. For the veterinarian whose goal is a healthy, fecund, self-sustaining colony, breeders that habitually produce nonviable or sick offspring are an economic and time-consuming liability. However, for many researchers such breeders and their offspring are a major resource for studying genetic, prenatal, and perinatal factors underlying both human and nonhuman primate obstetric and pediatric problems. Nevertheless, in most large primate facilities high-risk breeders and their offspring are systematically culled from the colony. This produces breeding groups that are increasingly free of naturally occurring attributes important for research on major human health problems. Within the economic conditions of large facilities such as Primate Research Centers, this unfortunate dilemma subverts the very goal of these institutions--namely, the maintenance of research colonies to solve scientific and health problems. It appears to us that this dilemma will be appreciated and solved only when administrators, veterinarians, and scientists wishing to exploit this resource become committed to a cooperative effort.

The beginning of such cooperation between administration, veterinary medicine, and medical-behavioral research is seen in interrelated programs at the Regional Primate Research Center and the Child Development and Mental Retardation Center at the University of Washington. Several grant-funded programs are involved in this effort. One studies respiratory disease in surgically delivered pigtail newborns. A second selectively breeds sires and dams at high or low risk for poor pregnancy outcomes (abortion, stillbirth, prematurity, low birth weight, neonatal death) and intensively studies physical, physiological, and behavioral development of surviving offspring. Other projects are concerned with sensory-motor development in high-risk and low-risk newborns.

Given this interest in research on high-risk monkeys, a collaborative program was developed by our administration, veterinarians, Infant Primate Research Laboratory staff, and researchers. Intensive medical care is given to neonates not assigned to research programs that are born prematurely, weigh less than normal, or are otherwise ill or traumatized while living with their mothers in the colony at the Primate Field Station of the Regional Primate Research Center. This "Save-A-Baby" program cares for high-risk neonates until they are no longer at risk medically. The infants are studied as they develop over the next 6-10 months, at which time they are returned to the colony for assignment to funded projects. Colony records showed that fewer than 5% of such high-risk newborns survived if left with their mothers, whereas survival rates in the nursery for this population now exceed 85%.

The benefits of this collaboration for both the colony and researchers include at least the following list. The program is economically sound, producing 10-15% more surviving infants per year. The invaluable normative and scientific developmental data collected by Infant Laboratory staff would be unavailable without this program, and the medical procedures developed for care of these high-risk neonates were critical for implementing a number of funded research projects.

BASIC HEALTH PARAMETERS IN THE NEWBORN NURSERY

The purpose of this chapter is to present some of the data generated through this "Save-A-Baby" program. These data measure development of four basic health parameters in normal and high-risk newborn Macaca nemestrina during the first month of life. Such information is crucial for understanding both normal and abnormal development of neonates. The chapter illustates the type of intensive data collection that is currently feasible only when an enlightened administration is willing and able to support basic research activities within the context of health care.

Perhaps the major health problem for both monkey and human high-risk neonates is regulation of basic physiological processes, including development of diurnal cyclicity (Bernard, 1973; Woodrum, Guthrie, and Hodson, 1977; Thoman, 1978). For premature humans and monkeys a primary cause of death is respiratory disease. Ability to regulate body temperature is often deviant for almost any type of high-risk neonate. Heart rate, as an index of autonomic nervous system functioning, is thought to be an important parameter in newborn health care (see Chapter 15). Central nervous system maturation is also reflected, at least in part, by development of normal sleep-wakefulness patterns. These parameters--sleep-wakefulness, respiration, temperature, heart rate--form a battery of information for health care and for assessing development of basic neural-physiological-behavioral mechanisms necessary for normal growth and maturation. Measurement of these parameters on a 24-hr per day basis has constituted an important activity of our Infant Laboratory over the past four years.

TABLE 1.   NUMBER OF INDIVIDUALS AND NUMBER OF ACTIVITY, HEART RATE, RESPIRATION, AND TERMPERATURE VALUES FOR STUDYING DEVELOPMENT OF BASIC HEALTH PARAMETERS IN NURSERY-REARED PIGTAIL INFANTS DURING THE FIRST 4 WEEKS OF LIFE.

Measure	Sex	Normal	Nursery Group Low Birth Weight	C- Section
Number of	Male	26	10	9
Individuals	Female	22	14	11
Number of	Male	6815	3765	2750
Activity Ratings	Female	5422	5130	3418
Number of	Male	1560	1055	947
Heart Rates	Female	1325	1470	1152
Number of	Male	1525	1020	955
Respiration Values	Female	1342	1440	1103
Number of	Male	1092	743	686
Temperatures	Female	928	1029	774

SUBJECTS AND MEASUREMENT PROCEDURES

Subjects and Housing

Sample sizes giving number of subjects and number of measurements are shown in Table 1. Subjects were 92 M. nemestrina neonates from three birth-condition groups, studied through the first 4 weeks of life. 1) Normal Nursery: 48 newborns were delivered naturally at full term. They were separated from their mothers at birth, or shortly thereafter. They were normal in birth weight and required no special medical care. 2) Low Birth Weight: 24 high-risk subjects, weighing less than normal at birth, were also delivered naturally. They were separated from their mothers within the first 1-3 days of life. The low-birth-weight criteria were derived from breeding colony norms (see Chapter 14) defining the 10th birth weight centile. This was at or below 405 g for males and 380 g for females. 3) Caesarean Section: 20 neonates were delivered by Caesarean section (C-section) at or near full term. These newborns were all within normal ranges for birth weight and were healthy at delivery.

Subjects lived in the Infant Laboratory nursery during days 1-28 after birth (see Chapter 13 for details of nursery procedures). Each newborn, regardless of health status, lived in an environment-controlled isolette during the first 7-10 days of life. On removal from the isolette, each animal was housed in a single cage in the same nursery room. Time from birth spent in isolettes averaged 16 days for low-birth-weight infants, 10 days for C-section monkeys, and 7 days for the normal group. Thus, some of our results may be influenced by differences in isolette time. However, all subjects did live in the same temperature-controlled nursery room during month one, and received a similar schedule of feeding, day-night light cycle (14 hr on/10 hr dimmed), and presence or absence of humans.

MEASURES

Sleep-wakefulness measurement was scheduled once per 30 min, 24 hr/day. On each visit an observer viewed the infant in its isolette or cage and rated it as being in one of four states: 1) behaviorally asleep, 2) awake but inactive, 3) awake and active, or 4) highly active and agitated. The fourth category rarely happened (less than 1% of total ratings), so its occurrences were pooled together with state 3. A complete 24-hr record thus contained 48 ratings, spaced at approximately 30-min intervals, for each subject. Data summaries for this activity level dimension are presented as percentages of occurrence for each state.

Respiration, heart rate, and temperature measurements were scheduled every 2-4 hr, 24 hr each day. The exact interval for each subject depended on need for health monitoring. Measurement proceeded by 1) observing the activity state, 2) visually counting chest wall movements (respiration) for 30 sec, 3) counting heart beats for 30 sec using a stethoscope, and 4) assessing rectal temperature using an electronic thermometer. The measurements thus were performed from the least manipulative to the most intrusive. The sequence was designed to minimize effects of one measurement upon the next in the series.

Two types of data presentations are made: Summaries by days-of-age and summaries by 4-hr time blocks of the 24-hr day. Sample calculations are illustrated in Table 2 for the heart rate and sleep state measures of a single subject.

TABLE 2.   ILLUSTRATION OF METHOD FOR CALCULATING DATA SUMMARIES BY 4-HOUR TIME BLOCK AND BY DAYS OF AGE FOR A SINGLE INFANT (explanation in text).

Calculation by 4-hr time block				Calculation by Days of Age			
Time Block 0-3				Days of Age 1-2			
Day of Age	Ratings	Sleeps	$\overline{X}$ HR	Time Block	Ratings	Sleeps	$\overline{X}$ HR
1	7	6	220	0-3	15	12	218
2	8	6	215	4-7	16	9	242
3	6	4	220	8-11	13	9	219
4	5	4	222	12-3	13	8	236
5	6	4	218	4-7	15	10	228
6	6	5	231	8-11	10	9	234
7	5	4	216				
Totals 43	33	1542		Totals 82	57	1377	
% Sleep	76.7			% Sleep	69.5		
Mean HR		220.3		Mean HR		229.5	

Days-of-Age

The number of times activity ratings were made in each of the six 4-hr daily time blocks was tallied. Also tallied were the number of times each activity state (sleep, awake-inactive, awake-active) occurred. These were summed over the six time blocks, and each state sum was divided by the total number of ratings made. For the sleep data illustrated on the right half of Table 2, this yielded a value of 69.5% sleep on days 1-2 of life. The basic physiological parameters (heart rate, respiration, temperature) were averaged for each of the six 4-hr daily time blocks. These values were then summed and divided by six to obtain the daily average. For the heart rate data in Table 2 this yielded an average rate of 229.5 beats/min on days 1-2 of life.

4-Hr Time Blocks

The left half of Table 2 illustrates calculation for the 4-hr time block 0-3, just after midnight to just before 4 a.m., in the first week of life. Total number of ratings in this time block for each day, number of occurrences of each activity state in this block, and mean heart rate, temperature, and respiration were summed over the 7 days of this week. The sum of each state was divided by 7. This yielded 76.7% as the value for sleep during week 1, and 220.3 beats/min as the average heart rate during week 1.

In the data presentation, time-of-day blocks always begin with the period just after midnight until just before 4 a.m. (time block 0-3). Day 1 began at 0 hr on the day the animal was born. Thus, for most animals day 1 did not include a full 24-hr period.

One complication concerns missing data. During most of the 4-yr period of acquiring these measures only one person was present on the late night/early morning shift, so if there were many animals in the nursery they could not all be tested in the time available. Usually only those infants requiring medical care were observed. Thus, the hours 11 p.m. to 5 a.m. contain fewer data values than most of the other hours of the day. Data were also missing when animals were removed from the nursery for behavioral testing. However, all subjects involved in the data presented here had at least one physiological value at each time-of-day, 10 physiological values for each day-of-age, and 20 total ratings for each percentage shown on the data graphs. Therefore, missing data probably produced little or no bias in either normative values or group effects.

Development of Sleep-Wakefulness Patterns

The basic diurnal activity pattern, averaged over all subjects for all 4 weeks, is shown in Figure 1. Sleep was cyclic, with peak periods from 8 p.m. to 4 a.m. Active and inactive wakefulness were not markedly differentiated, although the awake-and-active state was higher than the inactive state in the middle evening period when general laboratory activity is low.

Analyses of variance were performed assessing the main effects and interactions of birth condition, sex, weeks, and time of day with percentage of each activity state as dependent variables. The same analysis was performed with each of the three physiological measures as dependent variables. Statistically reliable sex differences appeared only for activity state main effects (all three p < 0.01). These effects are shown for data pooled over the first month of life in Figure 2. Females slept more than males, while males scored higher than females on each of the awake states. These basic sex effects appeared in each of the four test weeks.

The influence of birth condition on activity state during weeks 1-4 is shown in Figure 3. Low-birth-weight infants slept more than the other two groups, with C-section infants intermediate, and normal animals lowest. When awake, C-section animals

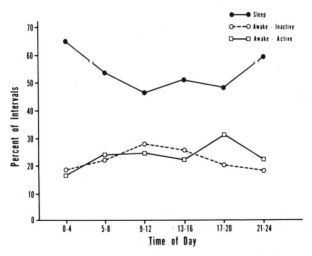

Figure 1. Overall diurnal activity of M. nemestrina newborns living in a nursery during the first 4 weeks of life.

were more inactive than low-birth-weight infants, while normal infants showed the least amount of inactive wakefulness. Active wakefulness was highest in normal animals, and did not differ between the other groups. These general effects are also seen in Figure 4 for each week. Thus, overall sleep was lowest and overall awake-and-active highest for the normal animals in each week. This would not seem to be an artifact of isolette living, as the normal and C-section infants did not differ markedly in time spent in that situation.

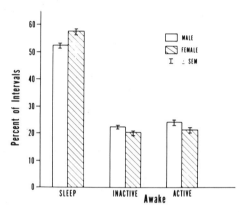

Figure 2.   Activity state differences between male and female M. nemestrina newborns during the first 4 weeks of life.

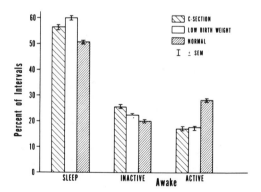

Figure 3.   Differences in activity state distributions during the first 4 weeks of life for normal, low-birth-weight, and C-section newborns.

Figure 4 also illustrates the development of diurnal cyclicity in each birth condition. Normal animals showed cyclic curves for each state beginning in week 1, and had markedly pronounced cycles for all states by week 3. Low-birth-weight and C-section infants did not show pronounced cyclicity in week 1, and their cyclicity was still not as strong by week 4 as that seen for normal infants. This suggests that normal animals begin regulating sleep-wakefulness rhythms right from birth and develop strong cyclicity sooner than either low-birth-weight or C-section neonates.

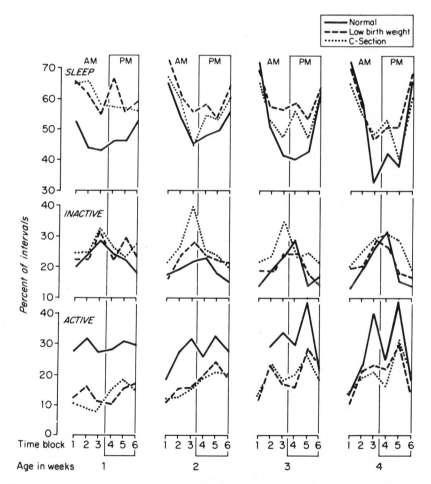

Figure 4.   Development of diurnal sleep-wakefulness cycles by normal, low-birth-weight, and C-section newborns.

Development of Basic Physiological Patterns

Figure 5 presents means by 2-day age blocks for temperature, respiration, and heart rate, contrasting normal versus low-birth-weight infants (A) and normal versus C-section infants (B).

In low-birth-weight infants, temperature was lower than that for normals for the first 16 days of life even though the low-birth-weight animals were in isolettes during most of this time. Respiration was initially lower than that for normals, but was much higher than normal by days 5-6. Heart rate was initially lower than

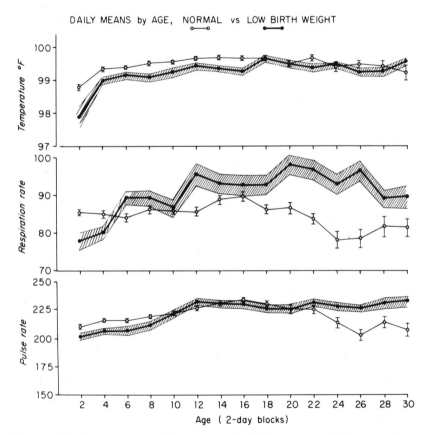

Figure 5A. Developmental changes in daily values of temperature, respiration, and heart rate for normal versus low-birth-weight newborn.

normal, did not differ from normal between days 10 and 20, then remained relatively high at ages when normal infant heart rates showed a marked decrease. In C-section infants, temperature was about the same as normal until day 26, when it rose to a higher level. Respiration was also about the same as for normals until day 20, when normal infants exhibited a marked decrease in their daily mean value. Heart rate was initially lower than normal, but became higher thereafter, although it followed the same basic pattern over days shown by normal infants.

These data reveal that development of basic physiological parameters over the first month of life clearly varies with birth

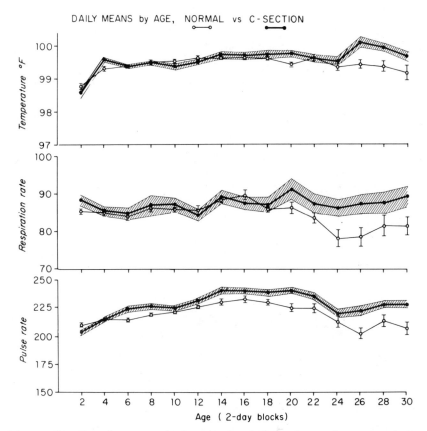

Figure 5B. Developmental changes in daily values of temperature, respiration, and heart rate for normal versus C-section newborns.

condition. As expected, low-birth-weight newborns differed considerably from normal on each measure. C-section animals showed
little or no initial differences from normal. Surprisingly, deviations from normal appeared later in month 1, resulting from
failure to drop in temperature or respiration averages during the
second 2-week period.

Figure 6 presents diurnal cycles for temperature, respiration,
and heart rate averaged over the first 4 weeks of life, contrasting

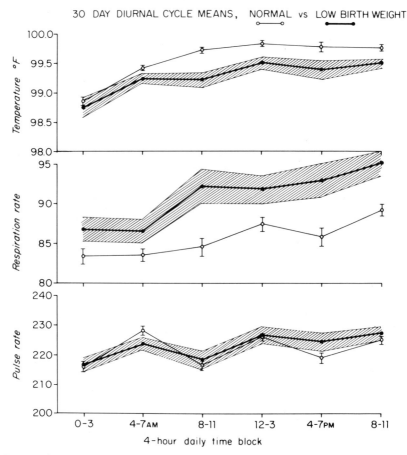

Figure 6A. Diurnal cycles of temperature, respiration, and heart
            rate during month one of life for normal versus low-
            birth-weight newborns.

low-birth-weight with normal infants (A) and C-section with normal infants (B).

Temperature cyclicity for all groups exhibited a gradual rise
throughout the 24-hr day, with the low point occurring during the
earliest morning hours.   Normal and low-birth-weight infants
showed a similar pattern, with temperature for the latter failing to
rise as high as that for normals.   Cyclicity in respiration differed
somewhat between normal and low-birth-weight infants.   The

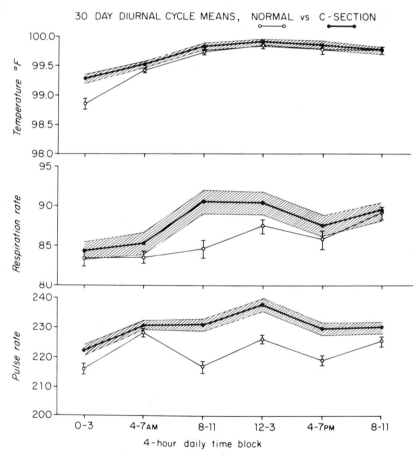

Figure 6B. Diurnal cycles of temperature, respiration, and heart
rate during month one of life for normal versus C-
section newborns.

latter showed a pronounced rise at time block 3, followed by relatively steady values thereafter. In normal infants, heart rate cyclicity followed an 8-hr pattern of low value/high value/low value/high value. Low-birth-weight infants showed this pattern in a less pronounced fashion during the morning hours, and did not show a heart rate drop at the 4-7 p.m. block. Overall temperature was higher for normals, while respiration was considerably lower.

The temperature patterns of C-section and normal infants were virtually identical, except that normal animals produced a much lower temperature during the 0-3 a.m. time block. For normals, respiration gradually increased throughout the 24-hr day, with a possible dip at the fifth block. C-section animals, like the low-birth-weight group, showed a different pattern with a marked rise between blocks 2 and 3, then a fall at block 7. Heart rate cyclicity was totally different for C-section infants; instead of showing a basic 8-hr cycle, heart rate gradually rose until evening, when it decreased. Overall respiration values were lower for normals, and overall heart rate was considerably lower for normal than for C-section infants.

Figure 7 presents weekly means illustrating development of diurnal cycles on each of the three physiological measures. Temperature cyclicity was present in all three groups for week 1. Over weeks, normal and C-section infants showed a similar pattern of cyclicity in temperature. Low-birth-weight animals had a deviant temperature pattern for weeks 2-3, and had markedly low temperatures during the first time block of weeks 1 and 4.

A similar respiration pattern was seen for all three groups in week 1, with group differences appearing in the time block containing peak values. In week 2, C-section animals had a different respiration cycle than the other two groups. In week 3, C-section and low-birth-weight patterns were somewhat different than normal. By week 4 all groups showed pronounced respiration cycles, with each group exhibiting at least some major difference in form of the pattern.

In week 1, normals showed little or no heart rate cyclicity, although the other groups did show changes over the six time blocks. By week 3, the normal animals had clearly established their basic 8-hr heart rate cycle seen in the overall data. The C-section group did not do so until week 4, and the low-birth-weight group never did. By week 4, heart rate and respiration cycles of normal animals were highly correlated (rose and fell together), a phenomenon that did not occur for the other two groups.

In general, these data suggest a relatively orderly set of changes for normal infants in both the form and amplitude of each

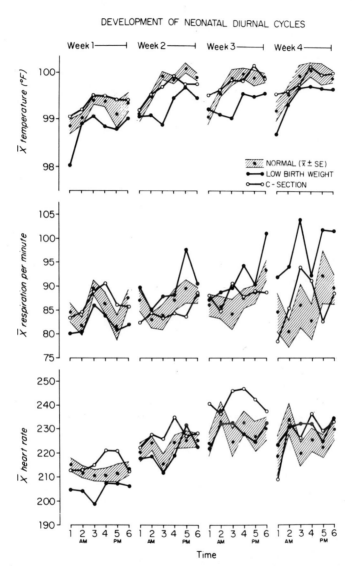

DEVELOPMENT OF NEONATAL DIURNAL CYCLES

Figure 7. Development of temperature, respiration, and heart rate diurnal cycles in normal, low-birth-weight, and C-section newborns.

diurnal function over the first 4 weeks of life. Except for temperature, this was not readily apparent in the other two groups. Low-birth-weight and C-section infants exhibited marked shifts in the shape of their diurnal functions and peak time periods from week to week.

CONCLUSIONS

The data presented here show that basic regulatory processes differ markedly with the newborn monkey's type of delivery and condition at birth. These birth variables affect both average daily values and diurnal cycles of activity states and physiological processes. Like high-risk human newborns (e.g., Illingworth, 1970; Metcalf, 1970), our low-birth-weight neonates appear to be deviant from the start of postnatal life. This reinforces our view that such animals can provide excellent models of human perinatal risk, and should therefore be considered a scientific asset in primate breeding programs.

Of special interest to us are the surprising effects found for C-section newborns. These animals, as a group, were apparently healthy at birth, were not premature or low in birth weight, and received almost the same schedule of postnatal conditions as their normal counterparts. Yet delayed delivery effects occurred, with marked differences between C-section and normal infants appearing 2-3 weeks after birth. This suggests that surgical delivery may pose health problems similar to those typically found for low-birth-weight and/or premature newborns.

The specific reasons for such C-section effects are not yet known. They could result from an abnormal hormone release or other mechanism normally initiated by natural delivery (Abuid, Sinson, and Larsen, 1973). Or, C-section delivery may produce infants who have difficulty adjusting to the transition from isolette to open-cage housing. The age of this transition (about 10 days) was just shortly before large differences appeared between normal and C-section infants. The problem warrants further investigation for at least two reasons: Many human practitioners believe that C-section delivery is especially safe for the neonate (although not necessarily for the mother); and some primate investigators routinely employ C-sections to minimize potential birth trauma effects in the interests of obtaining maximally normal newborns.

The basic normative data presented here have been used in our nursery as a major health monitoring tool. The data have also been used to assess developmental status and biological adaptive abilities of naturally occurring high-risk neonates and neonates receiving experimental treatments to themselves or to their mothers during pregnancy (Sackett, Holm, Davis, and Fahrenbruch, 1974; Sackett and Holm, 1978). The time, effort, and expense in gathering these data have thus been clearly worthwhile for us. To return to our earlier editorializing, we hope that administrators, veterinarians, and researchers will also see the value of these data. Possibly this could lead to coordinating the economic, technical, and physical resources needed to develop "Save-A-Baby" programs at other primate facilities.

ACKNOWLEDGMENTS

These studies were supported by NIH grants HD08633 and HD02274 from the NICHHD Mental Retardation Branch and RR00166 from the Animal Resources Branch. We thank Dr. Orville A. Smith, Director, Regional Primate Research Center at the University of Washington; Dr. Irvin Emanuel, Director, Child Development and Mental Retardation Center at the University of Washington; and Dr. Gerald A. Blakley, Supervisory Veterinarian, Medical Lake Breeding Colony. The work reported in this paper could not have proceeded without their active participation and administrative efforts in organizing our Infant Primate Research Laboratory facility.

REFERENCES

Abuid, J., Sinson, D. A., and Larsen, P. R. Serum triiodothyronine and thyroxine in the neonate and the acute increases in these hormones following delivery. J. clin. Invest. 52:1195-1199, 1973.

Barnard, K. The effect of stimulation on the sleep behavior of the premature infant. Commun. nurs. Res. 6:12-40, 1973.

Illingworth, R. S. Low birth weight and subsequent development. Pediatrics 45:335-339, 1970.

Metcalf, D. R. EEG sleep spindle ontogenesis. Neuropediatrics 1:428-433, 1970.

Sackett, G. P. and Holm, R. A. Effects of parental characteristics and prenatal factors on pregnancy outcomes of pigtail macaques. In: Maternal Influences and Early Behavior, R. W. Bell and W. P. Smotherman (Eds.), New York: Spectrum, 1978.

Sackett, G. P., Holm, R. A., Davis, A. E., and Fahrenbruch, C. E. Prematurity and low birth weight in pigtail macaques: Incidence, prediction, and effects on infant development. Pp. 189-205 in: Proceedings of the Fifth Congress of the International Primatological Society, Tokyo: Japan Science Press, 1974.

Thoman, E. B. The Origins of the Infant's Social Responsiveness. Hilldale, New Jersey: Lawrence Erlbaum, 1978.

Woodrum, D. E., Guthrie, R. D., and Hodson, W. A. Development of respiratory control mechanisms in the fetus and the newborn. In: The Development of the Lung, W. A. Hodson (Ed.), New York: Marcel Dekker, 1977.

III. HEALTH, DIET, AND GROWTH

CHAPTER 11

NURSERY REARING OF INFANT BABOONS

G. T. Moore and L. B. Cummins

Southwest Foundation for Research and Education
San Antonio, Texas

The infant baboon (Papio cynocephalus) has proven to be a useful biomedical research model and has been used in a variety of studies. These areas of investigation include nutrition, infectious diseases, toxicology of infant formula preservatives, and diet as it affects the development of atherosclerosis. This increased use of the infant baboon for research has precipitated a need for more expertise in rearing the infants in a nursery environment and collecting baseline data on basic care and normal physiology of the animals. Approximately 100 baboon infants have been successfully reared in a conventional nursery at the Southwest Foundation for Research and Education (SFRE) during the past three years. These animals were offspring of both P. c. anubis and P. c. cynocephalus. Births occurred throughout the year with no significant birth season established. The lack of a birth season has been observed in captive baboons once they acclimated to their environment.

General nursery management principles for baboons have been previously discussed (Miller and Pallotta, 1965; Vice, Britton, Ratner, and Kalter, 1966). Many of the guidelines established by these reports are followed in rearing baboons at SFRE, but some have been altered to accommodate special experimental needs.

METHODS

Husbandry practices and the breeding management for the baboon colonies at SFRE have been reported elsewhere (Kraemer and Vera Cruz, 1972; Moore, 1975; Kraemer, Kalter, and Moore, 1975). Most of the baboon infants are born in the outdoor breeding groups, but some are born indoors to individually caged females. Females that deliver in the outdoor group cages are allowed to go undisturbed to term and the infant is separated from its mother as soon as it is observed. In established social groups there is little fetal waste from traumatic incidents. The infant in these conditions is almost always 6-12 hr old upon separation and has had an opportunity to drink colostrum. When a situation dictates that a female deliver indoors in an individual cage, she must be established in that environment at least 3-4 weeks before parturition. Moving the pregnant female immediately before parturition many times results in the mother either killing her infant or rejecting it to die of neglect.

All infants born indoors that are scheduled to be nursery reared are allowed to remain with their mothers 6-8 hr after birth, giving them time to nurse colostrum. This also allows time for the mother to clean the infant, remove the placenta, and sever the umbilical cord.

Mothers from the outdoor group environments are sedated with ketamine hydrochloride (10 mg/kg) given intramuscularly in order to safely remove the infants. Most infants from mothers housed in individual cages that have squeeze mechanisms can be separated without sedating the mother, but sometimes an infant is injured seriously in the process. The separated infant is dried thoroughly with a sterile towel and the umbilical cord is ligated with silk and severed one inch from the abdominal wall. The severed end of the umbilical cord is treated with iodine and the infant is moved to the nursery. A warm bath with dilute Betadine surgical soap is given the infant if it is soiled with any dirt or fecal material.

Before admission into the nursery each infant is given a thorough physical examination. All nursery personnel in direct contact with the infant must wear surgical gowns, gloves, masks, caps, and shoe covers while working with the infants. The number of personnel attending the nursery infants is kept to a minimum and no one else is allowed to enter the nursery except for pertinent business. The physical examination includes checking heart rate and sound, respiratory rate and character, appearance of mucous membranes, rectal body temperature and body weight, and looking for anomalies such as atresi ani. This information is recorded and copies are submitted to the computer records. After the physical examination a complete blood count is taken, plus oral and rectal swabs for monitoring intestinal microbiological flora, then the infant is placed in an isolette. The principal need for the isolette is to give the infant's thermoregulating system time to adjust. This generally requires 2-3 days in most animals. An isolette temperature of 88-92° F provides best results. The infant's rectal temperature will average 98° F in these conditions. Diapers are not used as a rule, but a towel is placed in the isolette with the infant and changed after each feeding.

The first feeding begins 12-18 hr after birth with the formula to be used started immediately in lieu of 5% dextrose or saline and dextrose. Standard 4-oz formula bottles and pre-packaged nipples are used. The standard human infant nipples need to be softened by boiling or autoclaving in water before use, and the nipple holes enlarged slightly to provide easier delivery of formula needed to promote adequate growth. If the holes are not enlarged, the baboon infant quickly tires of sucking and will not consume the necessary formula. Unused formula is discarded after each feeding and all nipples that are reused are sterilized between feedings. The standard nursery feeding schedule at SFRE is shown in Table 1. The formula is fed in bowls beginning at 85 days and baboon chow is added to formula to make a mash at 98 days.

The most commonly used formulas have been commerical products for humans, Enfamil (Mead-Johnson) and Similac (Ross Laboratories). These formulas contain 20-26 kcal/fluid oz, and can be fed in the concentrated form to increase calorie intake. Infants have been reared successfully on diets that provide 120-258 kcal/kg/day. The higher caloric diets generally resulted in better weight gains and more healthy animals with fewer clinical disease problems.

Each animal is weighed daily before the morning feeding to obtain a normal fasted weight. Next, a general physical examination is made to determine rectal temperature, stool condition, and general activity level. Feeding is next, with the formula bottle and nipple weighed before and after to determine grams of diet consumed. "Burping" the infant is very important to reduce vomiting and possible aspiration of milk, but should be done only after it has consumed its normal amount of diet, as premature burping may result in reduced suckling activity. All infants are hand-fed for at least one month to insure an adequate formula intake. Self-feeding is started at 4-6 weeks based on the progress of the individual animal. Self-feeding in the baboon could start earlier, but formula is wasted from dripping and inadequate sucking by the infant, which would interfere with adequate monitoring of formula consumption on nutrition studies. Other reports (Vice et al., 1966) have indicated that baboon infants can be started on self-feeding as early as 8-20 days of age.

The infants are moved from the isolettes into hanging wire cages by 10-14 days of age. These cages allow more activity and

TABLE 1. NURSERY FEEDING SCHEDULE FOR BABOONS

Infant Age (Days)	Feedings/ Day	Comment
1-14	5	7 and 11 a.m., 3, 7, 11 p.m.
15-28	4	7 and 11 a.m., 3 and 7 p.m.
29-84	3	7 and 11 a.m., 3 p.m.
85-98	2	7 a.m., 3 p.m.; formula in bowl
98-112	2	7 a.m., 3 p.m.; formula in bowl with baboon biscuits.

provide a cleaner environment. The necessity of housing the infants separately for nutrition studies has resulted in the development of some abnormal characteristics. Some animals curl up in the bottom of the cage and show reduced activity, while others suck on penis, thumb, or toe. Providing towels as security blankets and hanging the towels from the top of the cage has helped to prevent the animals from "balling up" in the bottom of the cage. In addition, beginning at one week of age, infants were placed in an activity cage with one or two peers for 2 hr a day. We have observed no behavioral problems since making these changes.

A good sanitation program is an absolute necessity in a baboon nursery. The isolettes are cleaned and disinfected daily with a phenolic-based compound, O-SYL (National Laboratories). The wire caging is cleaned once weekly in a commercial cage-washing machine using a sanitizing hot water cycle at 185° F.

Complete physical examinations are made on each infant monthly; this includes a complete blood count, stool sample examination for parasitology, and oral and rectal swabs for monitoring intestinal microbiological flora. All infants are observed daily for signs of overt clinical illness with pneumonia and intestinal infections noted most often. The infant baboon responds very well to treatment and the death rate in the SFRE nursery has been less than 5%. Deaths are predominately due to pneumonia and birth anomalies. Intestinal infections rarely cause death in treated baboon infants. Atresia ani, hydrocephalus, and blindness have been the most common birth defects observed.

RESULTS AND DISCUSSION

Two groups of baboons reared at different times in the SFRE nursery were evaluated. Group I consisted of 9 males on an atherosclerosis-nutrition study. These animals were fed Enfamil, which provided 20 cal/oz. The mean weekly formula consumption of this group is shown in Figure 1. There was a mean formula consumption of 43 oz/infant the first week and 72 oz/infant the 14th week. This provided a mean weekly intake of 64.4 oz and a daily mean intake of 9.2 oz. The mean daily caloric intake of this formula was 184 kcal/baboon. The total caloric consumption ranged from 120 to 146 kcal/kg/day, with a mean of 124 kcal/kg/day. The mean weekly weights for this group are shown in Figure 2. The mean weight of the group at the end of one week of age was 1,026 g. The mean weight gain for the 13 weeks in the nursery was 634 g. This was a mean daily gain of 6.96 g and a mean weekly gain of 48.76 g for the group. The most rapid weight gains occurred in the first 6 weeks, with a reduction in the rate of gain during weeks 7-11. The rate of gain did increase again after week 11 to show a continuous increase of weight during the 13-week period.

A second group of 24 infants (9 males and 15 females) was studied to determine the effects of an infant formula preservative. Their diet provided 20 kcal/oz, which is similar to that of the first group studied. The estimated weight percentiles for this group have been previously reported (McMahan, Wigodsky, and Moore,

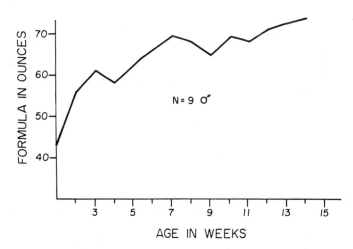

Figure 1.   Mean weekly formula consumption in baboons.

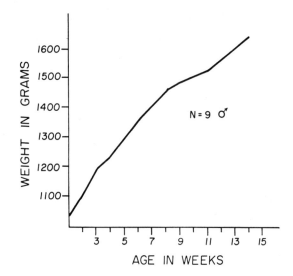

Figure 2.   Mean weekly weight gain in baboons.

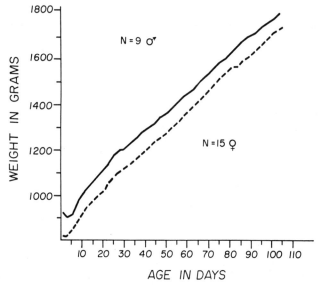

Figure 3.   Mean daily weights in baboons.

1976). The initial weights were collected 18-24 hr after birth and thus are not a true birth weight. The weights were recorded within 24 hr of birth and thereafter each morning before the first feeding. Figure 3 shows the mean daily weights of this group. The mean initial weight was 920 g for the males and 820 g for the females. The males showed a drop of about 15 g from the first weight in the second and third days. The females showed a weight loss also in the second and third days, but only an average of 3 g. The males gained an average of 887 g during the 105-day period with a daily mean weight gain of 8.4 g. The females gained an average of 909 g during a comparable period for a daily mean weight gain of 8.6 g.

The methods used in this study have proven to be successful in rearing the infant baboon in a nursery facility at SFRE. A previous report (Vice et al., 1966) indicates that a diet providing 26.7 kcal/oz was more satisfactory than one that offered 13.2 kcal/oz. The present report demonstrates that a diet with 20 kcal/oz has resulted in infants growing faster and weighing more at weaning than the higher caloric diet of 26.7 kcal/oz. It seems that a formula containing approximately 20 kcal/oz when fed to baboon infants in a well-managed nursery results in excellent weight gains and healthier infants.

REFERENCES

Kraemer, D. C., Kalter, S. S., and Moore, G. T. The establishment of non-human primate breeding colonies at the Southwest Foundation for Research and Education. Lab. anim. Sci. 6:41-46, 1975.

Kraemer, D. C. and Vera Cruz, N. C. Breeding baboons for laboratory use. Pp. 42-47 in: International Symposium on Breeding Non-human Primates for Laboratory Use, W. I. B. Beveridge (Ed.), Basel: Karger, 1972.

McMahan, C. A., Wigodsky, H. S., and Moore, G. T. Weight of the infant baboon (Papio cynocephalus) from birth to fifteen weeks. Lab. anim. Sci. 26:928-931, 1976.

Miller, R. L. and Pallotta, A. J. Comments on maintenance of a small baboon colony. Pp. 111-124 in: The Baboon in Medical Research, Proceedings of the 1st International Symposium on the Baboon and Its Use as an Experimental Animal. San Antonio: University of Texas Press, 1965.

Moore G. T. The breeding and utilization of baboons for biomedical research. Lab. anim. Sci. 25:798-801, 1975.

Vice, T. E., Britton, H. A., Ratner, I. A., and Kalter, S. S. Care and raising of newborn baboons. Lab. anim. Sci. 16:12-22, 1966.

CHAPTER 12

GROWTH AND DEVELOPMENT OF INFANT SQUIRREL
MONKEYS DURING THE FIRST SIX MONTHS OF LIFE

J. N. Kaplan

Developmental Psychobiology Program
SRI International
Menlo Park, California

INTRODUCTION

The experiences of several laboratories over the last few years have demonstrated that the squirrel monkey (Saimiri sciureus) can be bred under a variety of captive conditions and that the offspring can be maintained with little difficulty (Hayes, Fay, Roach and Stare, 1972; Rosenblum, 1972; Taub, Adams and Auerbach, 1977; Kaplan, 1977c). This is particularly significant for the investigator who is interested in reproductive and developmental processes of primates but limited in terms of space and budget. The small size and relatively tractable nature of the squirrel monkey make it an excellent subject for many research objectives.

Normative developmental data, including parameters of physical growth, dental eruption, hematology, and behavior have been reported for Saimiri (Long and Cooper, 1968; Rosenblum, 1968a; Ploog, 1969; Kaplan, 1974, 1977b; Ausman, Gallina, Hayes, and Hegsted, 1976). There have also been reports on general procedures covering natural and artificial rearing (Ausman, Hayes, Lage, and Hegsted, 1970; Hinkle and Session, 1972; Garcia, 1976; Kaplan, 1974, 1977c). However, in almost all cases the results presented for a specific parameter (e.g., hematology) have been based on only one of the several "subtypes" of Saimiri, largely because most captive breeding colonies of Saimiri include only the type that was most abundant from a particular supplier at the time the colony was initially established. Prior to 1974, the two major cities from which Saimiri were exported were Leticia, Colombia, and Iquitos,

Peru.   Animals from these two regions are distinctly different (Cooper, 1968; Jones and Ma, 1975), and most of the existing captive breeding colonies consist of one or the other of these two varieties.   Peru and Colombia stopped exporting nonhuman primates in 1975. Bolivia, which had exported only a nominal number of monkeys before 1974, then became the major source of Saimiri. Saimiri from Bolivia are phenotypically different than those from both Colombia and Peru, but because of their relatively recent introduction into the laboratory, detailed comparisons with the other two types have not yet been made.

Another limitation of much of the developmental research with Saimiri, particularly in relation to those studies that have concentrated on physical or physiological parameters, is that differences in rearing techniques (e.g., rearing with mother vs. artificial rearing) generally have not been taken into account. This does not mean that such differences are likely to produce contradictory results, but, rather, that evidence is needed to discount this factor as a source of variation.

This chapter briefly describes some of the features of physical and behavioral development of the squirrel monkey during the first 6 months of life, along with the methods of rearing routinely used in our laboratory.   The data are based on infants born in our breeding colony and used primarily for studies dealing with behavioral development.   Initially, our colony consisted of only the Peruvian type of Saimiri, but it gradually expanded to include an even larger number of the Colombian variety. A few years ago we began to acquire monkeys from Bolivia as well, so that we now have obtained certain comparative data on infants of the three subtypes.   More than 50 viable offspring have been born in the colony in each of the last several years.

In addition to making comparisons among these three subtypes, we have also been able to assess some of the consequences of natural and artificial rearing procedures.   In particular, comparisons have been made between infants raised with their natural mothers and those raised with a fabricated, inanimate surrogate mother. The surrogate used in our laboratory has been described in detail elsewhere (Kaplan and Russell, 1973). Essentially, it consists of a heated plastic cylinder approximately the size of an adult female; it contains a bottle for nursing and has an outer fur-like cover for comfort and clinging. The temperature of the surrogate is maintained at approximately $34^{\circ}C$.   Although we have never directly evaluated the necessity of heating the surrogate, infants have shown clear signs of disturbance on occasions when a surrogate's heating system has malfunctioned temporarily.

GROWTH

Surrogate-reared infants in our laboratory are fed a standard commercial milk-substitute designed for human infants.   After minimal training, they are capable of nursing by themselves within the first 2 weeks of life. The milk formula is provided throughout the first 6 months, but the infant's diet is extended to include commercially prepared baby cereal during the period of approximately 4-8 weeks of age, water-softened monkey biscuits beginning at 6 weeks of age, and a variety of fruits that can be easily consumed and digested beginning at about 6 weeks. The growth and physical development of infants reared in this manner have been virtually identical to those of infants reared by natural mothers in our laboratory. Figures 1 and 2 show increases in weight during the first 24 weeks of life for mother- and surrogate-reared infants, respectively. Each figure presents separate curves for males and females of the different types. Although there is no difference between the two rearing conditions, clear differences are found between the different subtypes under each condition. The largest contrast is found between the Colombian and Peruvian animals. Colombian infants are consistently heavier than Peruvians from birth, and the degree of magnitude increases with age. Bolivian infants tend to fall between the other two varieties, although by 24 weeks mother-reared Bolivians weigh as much as Colombian animals. Mother-reared males of each of the three types tend to be heavier than their female counterparts, beginning after approximately 12 weeks of age, although this difference is not nearly as great as that found between the different types and is less prominent in surrogate-reared subjects.

HEALTH

The most suitable procedures for raising infants depend on the objectives of the research in which they are being used. The easiest and least expensive method is to leave infants with their mothers until they can be readily weaned, around 4-6 months of age. However, we have found that infants reared in such a manner have a higher mortality rate before weaning than do those reared by artificial means. One possible reason for this is that subtle changes in an infant's health are less easily detected when it is living with its mother and may not be noticed until it is too late to provide adequate treatment. Also, problems attributable to a mother's inability or unwillingness to provide the necessary care and to interference by other group members under conditions of group housing occasionally add to the increased incidence of deaths.

The majority of infant deaths in our laboratory have occurred quite unexpectedly, without any previous indication of illness, and seem to have been ultimately precipitated by severe dehydration. Saimiri infants have a paucity of body fat and seem to be particularly sensitive to fluid imbalance. Approximately 15% of our mother-reared infants have died before the age of 6 months. The largest proportion of these deaths have occurred in the first

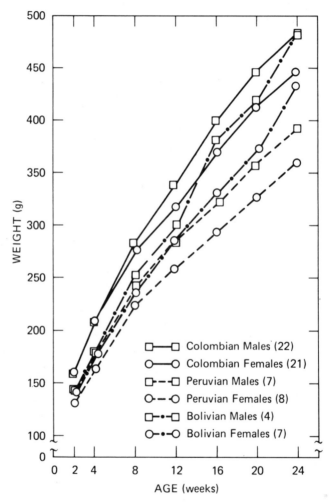

Figure 1.   Weights of mother-reared infant squirrel monkeys of three subtypes.

week, followed by a high percentage between 4 and 12 weeks of age. The reason for the high incidence of deaths in the latter period is not readily apparent, but one possibility is that infants in this age range may be particularly susceptible to certain infectious agents from which they have been previously protected by maternal antibodies acquired in utero. In contrast, there have never

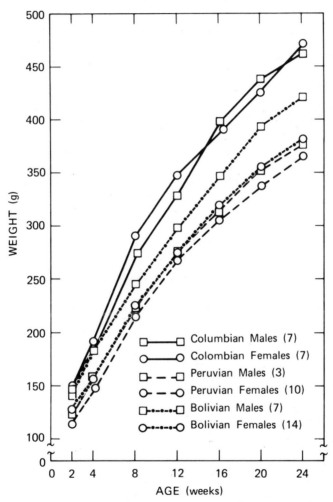

Figure 2.   Weights of surrogate-reared infant squirrel monkeys of three subtypes.

been any unexplained deaths of artificially reared infants in our laboratory, who otherwise have lived in environments that are quite comparable to those for mother-reared animals. Therefore, it is possible that if an infectious agent is involved, it may be transmitted directly by the mother, perhaps through her milk. Demographic data covering the 1975, 1976, and 1977 seasons in our laboratory for mother-reared and surrogate-reared infants combined are presented in Table 1. Surrogate-reared infants included 7 Colombians, 10 Peruvians, and 24 Bolivians of which only 3 died before 24 weeks of age, all from identifiable causes. It is interesting to note that of the mother-reared animals, all but one of the infants that died between the ages of 4 and 12 weeks were Colombian. Although we do not believe that this period of apparent increased sensitivity is limited to Colombian infants, since in past years our Peruvian infants have also died during this time, we do feel that whatever might be causing the disproportionate number of deaths during this period has a greater effect on the Colombian type.

In general, the first indication that an infant is weak and dehydrated is its complete inability to maintain itself on its mother's back. Mothers often try to support and carry infants that

TABLE 1. NUMBER OF LIVE BIRTHS AND DEATHS (TO 24 WK OF AGE) OF THREE TYPES OF Saimiri BORN AT SRI INTERNATIONAL IN 1975, 1976, and 1977

Live births	Colombian	Peruvian	Bolivian
1977	34	11	24
1976	28	14	12
1975	35	18	1
Total	97	43	37
Deaths			
Less than 1 wk old	6	3	4
1-4 wk old	1	2	0
4-12 wk old	11	0	1
12-24 wk old	2	1	1
Total	20 (21%)	6 (14%)	6 (16%)

have lost this ability, but this does not provide any help other than possibly temporarily keeping infants warmer and less stressed than if they were left on the ground. Infants found in this condition often exhibit seizures. We typically administer replacement fluids subcutaneously at regular intervals to weakened infants and keep them in a nursery incubator. Water with dextrose and/or other liquids are administered in small quantities by mouth as soon as infants will ingest them voluntarily and tolerate them without regurgitation. Unfortunately, only a small percentage of infants younger than 8 weeks have survived in our laboratory after developing these symptoms.

BEHAVIOR

Our experience has indicated that much of the behavior of mother- and surrogate-reared animals in the first 6 months of life is very similar, even though the degree of stimulation is undoubtedly greater in progeny living with real mothers. However, infants raised on inanimate surrogates or by other artifical procedures frequently develop thumbsucking behavior (which in some cases in our laboratory has persisted until after 3 years of age) and occasionally display other forms of unusual self-directed responses (e.g., grabbing and pulling at parts of their bodies). Neither of these types of behavior has ever been observed in any of our mother-reared animals. Moreover, we have also found that at the time of weaning, infants reared alone with only a surrogate mother may have difficulty adjusting to new surroundings that include agemates--especially those reared by natural mothers. Under such conditions, previously healthy infants have stopped eating and drinking and have shown other signs of depression. In contrast to these rearing-related differences, which--it should be emphasized-- are not nearly as severe as those reported for certain Old World species (Mitchell, 1970), we have not found any differences in the development of certain reflex-like or other simple behaviors. For example, surrogate-reared infants in our laboratory have performed as well as those reared by their natural mothers on a variety of tests designed to monitor the development of such responses as grasping, climbing, body orientation, locomotion, and sensory perception.

Infants reared with an inanimate surrogate mother develop a very strong attachment to their maternal figure, comparable to the attachment shown by mother-reared infants to their mothers. In addition to observing this at the behavioral level (Kaplan, 1977a, 1977b), we have recently compared pituitary-adrenal changes in mother- and surrogate-reared infants exposed to the stress of physical separation from their respective maternal figures when infants were approximately 3 months old (Mendoza, Smotherman, Miner, Kaplan, and Levine, 1978). Following a 30-min separation, increases over baseline values in levels of plasma cortisol were similar for infants reared in each of the two rearing conditions, suggesting similar strengths of attachment to inanimate and animate maternal objects.

Although infants with different rearing experiences respond similarly in terms of their simple behavior patterns, certain basic differences among the three subtypes have been observed. We have generally found Peruvian infants to be more emotional than either the Colombian or Bolivian types. For example, Peruvian infants appear more fearful in stressful situations (e.g., novel environments) and tend to "freeze" and emit stress-related vocalizations to a greater extent than do either of the other two varieties. Bolivian infants have appeared to be the calmest of the three types under a variety of test conditions.

These differences are evident to some extent in a simple test we have used to measure changes in the development of climbing during the first 8 weeks of life. In this test, infants are placed at the bottom of a rectangular sheet of wire mesh attached to a 1.3-m household ladder at a 70° incline. The mesh is marked off into seven consecutively numbered 15-cm sections and the distance climbed (i.e., highest section reached) within 30 sec is scored. As depicted in Figure 3, climbing distances for all three subtypes decrease with age, owing to a decrement in the strength of the reflexive component of climbing evident at early ages. However, the pattern of change over the 8-week test period is clearly different for the different types. These differences are best explained by the way the three types of infants behave in the test situation, particularly in the first few weeks, which would also argue against an interpretation based on differences in climbing ability per se. Throughout most of the 8-week test period, Peruvian infants typically remained immobile and vocalized during trials, which clearly was incompatible with climbing behavior. Many of the Colombian infants reacted in a similar fashion during the first weekly test, but thereafter apeared much more relaxed. Bolivian infants both "froze" and vocalized the least of the three types from the beginning of testing and, in contrast to the other types, did not hesitate to climb the mesh incline in the first week.

Along similar lines, we have also found that among infants raised on inanimate surrogates, Colombian infants show a more rapid decline than Peruvian infants do in the amount of time spent

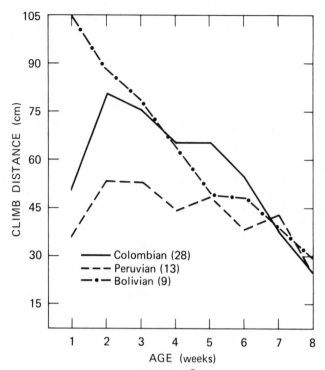

Figure 3.   Distance traveled up a 70° incline during 8 weekly 30-sec tests for infant squirrel monkeys of three subtypes.

on their rearing surrogate (Kaplan, 1974) during the first 24 weeks of life. Like the results presented above, this could be interpreted in terms of the seemingly more fearful nature of infants of the Peruvian subtype.

In all the studies we have conducted to date, very few behavioral differences have been found between males and females younger than 6 months of age. One conspicuous difference we have observed, however, concerns the genital display, a reflexive type of behavior that is characteristic of squirrel monkeys, and in adults is performed more often by males than by females (Ploog, 1967). Although the response is basically similar for adults of both sexes, adult females occasionally display by simultaneously spreading both legs outward, instead of the more common movement of abducting only one leg. Such postural dimorphism is seen more clearly in immature animals, and we have studied it to some extent in our laboratory by exposing infants 4-6 months of age to their mirror-image, a condition that tends to elicit such displays. In brief, these tests have revealed that males of this age respond to their mirror image predominantly in the one-legged fashion, whereas females respond most often in the two-legged manner. This occurred regardless of whether infants were raised with natural mothers or artificial surrogate mothers.

SUMMARY

It seems clear that the popularity of the squirrel monkey as a laboratory animal will continue to increase as the costs of maintaining larger primates increase. In light of the current trend of curtailing supplies of nonhuman primates from their native habitats, it is reassuring to know that the initial problems enountered in breeding and raising squirrel monkeys in captivity (Goss, Popejoy, Fusiler, and Smith, 1968; Rosenblum, 1968b) have largely been eliminated through a better understanding of their husbandry and reproductive physiology. Researchers interested in using a nonhuman primate to study developmental processes now have the opportunity to establish and maintain their own production facility at substantially lower costs than would be necessary with most other primates.

ACKNOWLEDGMENTS

Supported by NIH grant HD04905 from the National Institute of Child Health and Human Development. The assistance of C. McKenzie and D. Odlin is gratefully acknowledged.

REFERENCES

Ausman, L. M., Gallina, D. L., Hayes, K. C., and Hegsted, D. M. Hematological development of the infant squirrel monkey (Saimiri sciureus). Folia primat. 26: 292-300, 1976.

Ausman, L. M., Hayes, K. C., Lage, A., and Hegsted, D. M. Nursery care and growth of Old and New World infant monkeys. Lab anim. Care 20: 907-913, 1970.

Cooper, R. W. Squirrel monkey taxonomy and supply. In: The Squirrel Monkey, L. A. Rosenblum and R. W. Cooper (Eds.), New York: Academic Press, 1968.

Garcia, F. G. Primate husbandry and reproduction. In: First Inter-American Conference on Conservation and Utilization of American Nonhuman Primates in Biomedical Research. Wash., D. C.: Sc. Pub. No. 317, Pan American Health Org., 1976.

Goss, C. M., Popejoy II, L. T., Fusiler, J. L., and Smith, T. M. Observations on the relationship between embryological development, time of conception, and gestation. In: The Squirrel Monkey, L. A. Rosenblum and R. W. Cooper (Eds.), New York: Academic Press, 1968.

Hayes, K. C., Fay, G., Roach, A., and Stare, F. J. Breeding New World monkeys in a laboratory environment. In: Breeding Primates, W. I. B. Beveridge (Ed.), Basel: S. Karger, 1972.

Hinkle, D. K. and Session, H. L. A method for hand rearing of Saimiri sciureus. Lab. anim. Sci. 22:207-209, 1972.

Jones, T. C. and Ma, N. S. F. Cytogenetics of the squirrel monkey (Saimiri sciureus). Fed. Proc. 34:1646-1650, 1975.

Kaplan, J. Growth and behavior of surrogate-reared squirrel monkeys. Develop. Psychobiol. 7:7-13, 1974.

Kaplan, J. Perceptual properties of attachment in surrogate-reared and mother-reared squirrel monkeys. In: Primate Bio-social Development, S. Chevalier-Skolnikoff and F. E. Poirier (Eds.), New York: Garland, 1977a.

Kaplan, J. Some behavioral observations of surrogate-and mother-reared squirrel monkeys. In: Primate Bio-social Development, S.Chevalier-Skolnikoff and F. E. Poirier (Eds.), New York: Garland, 1977b.

Kaplan, J. Breeding and rearing squirrel monkeys (Saimiri sciureus) in captivity. Lab. anim. Sci. 24:557-567, 1977c.

Kaplan, J. and Russell, M. A surrogate for rearing infant squirrel monkeys. Behav. res. meth. Instru. 5:379-380, 1973.

Long, J. O., and Cooper, R. W. Physical growth and dental eruption in captive-bred squirrel monkeys, Saimiri sciureus. In: The Squirrel Monkey, L. A. Rosenblum and R. W. Cooper (Eds.), New York: Academic Press, 1968.

Mendoza, S. P., Smotherman, W. P., Miner, M. T., Kaplan, J., and Levine, S. Pituitary-adrenal response to separation in mother and infant squirrel monkeys. Develop. Psychobiol., 11:169-175. 1978.

Mitchell, G. Abnormal behavior in primates. In: Primate Behavior: Developments in Field and Laboratory Research (Vol. 1), L. A. Rosenblum (Ed.), New York: Academic Press, 1970.

Ploog, D. W. The behavior of squirrel monkey (Saimiri sciureus) as revealed by sociometry, bioacoustics, and brain stimulation. In: Social Communication Among Primates, S. A. Altmann (Ed.), Chicago: Univ. Chicago Press, 1967.

Ploog, D. W. Early communication processes in squirrel monkeys. In: Brain and Early Behavior, R. J. Robinson (Ed.), New York: Academic Press, 1969.

Rosenblum, L. A.   Mother-infant relations and early behavioral
    development in the squirrel monkey.   In: The Squirrel
    Monkey, L. A. Rosenblum and R. W. Cooper (Eds.), New York:
    Academic Press, 1968a.
Rosenblum, L. A.   Some aspects of female reproductive physiology
    in the squirrel monkey.   In: The Squirrel Monkey, L. A.
    Rosenblum and R. W. Cooper (Eds.), New York: Academic
    Press, 1968b.
Rosenblum, L. A.   Reproduction of squirrel monkeys in the
    laboratory.   In: Breeding Primates, W. I. B. Beveridge (Ed.),
    Basel: S. Karger, 1972.
Taub, D. M., Adams, M., and Auerbach, K. G.   Reproductive
    behavior in a breeding colony of Brazilian squirrel monkeys
    (Saimiri sciureus).   Paper presented at the meeting of the
    American Society of Primatologists, Seattle, April 1977.

CHAPTER 13

SURVEY OF PROTOCOLS FOR NURSERY-REARING INFANT MACAQUES

G. C. Ruppenthal

Infant Primate Research Laboratory
Child Development and Mental Retardation Center
and Regional Primate Research Center
University of Washington
Seattle, Washington

INTRODUCTION

If there is one monkey species that could be considered the standard laboratory animal for use in biomedical and behavioral research, it would have to be the rhesus monkey (Macaca mulatta). This species, known for its hardiness and adaptability to the rigors of laboratory confinement, has been the mainstay of experimental research for many years. Consequently, a wealth of data on many parameters of growth and development exists for this species. Many nursery facilities throughout the world have continued, modified, or adopted the techniques for rearing nonhuman primates from procedures initiated by a few individuals years ago for rhesus monkeys.

Another species that has recently come into the forefront as a research animal is the pigtail macaque (M. nemestrina). Similar in many respects to the rhesus macaque but with some important differences, such as a brain-to-body size ratio more similar to that in humans, the pigtail macaque also adjusts well to laboratory regimen and reproduces well in captivity. Other macaque species used in several facilities include the bonnet (M. radiata), the crabeating or long-tailed (M. fascicularis), the stumptail (M. arctoides), and the Japanese macaque (M. fuscata). All of these species do well in captivity, and investigators rearing them owe at least some of their success to a combination of facts and lore first accumulated by individuals working with rhesus monkeys.

165

This chapter is a survey of protocols used in several large nursery facilities in the United States. The majority of the procedures discussed are practiced in our laboratory; however, we do not mean to promote them as "the way to do it." They are offered as guidelines that we have found to be sufficient for our own needs. Indeed, macaques are extremely adaptable and can be reared under a vast array of protocols. Research emphasis, reproductive rate, need for socialization, and cost must be considered before nursery rearing practices can be determined.

The data reported here come from four facilities that maintain primate nurseries: the University of Wisconsin Department of Psychology Primate Laboratory, which is supported in part by the Wisconsin Regional Primate Research Center; the Oregon Regional Primate Research Center; the Hazleton Laboratories America, Incorporated; and the Infant Primate Research Laboratory which is a core facility of the Child Development and Mental Retardation Center and the Regional Primate Research Center at the University of Washington. The Washington Primate Center also maintains the Primate Field Station, a breeding and research facility at Medical Lake, WA, which supplies many of the neonates in the Infant Laboratory.

PRENATAL CARE AND PREGNANCY OUTCOME

Diet

Hazleton, Oregon, Washington, and Wisconsin all currently feed commercial monkey chow (Ralston-Purina Co., St. Louis, MI) as the primary source of nutrition for females, with no special provisions added during pregnancy. Amounts and supplements are listed in Table 1.

Fetal Loss

Many factors have been shown to influence pregnancy outcome in macaques. For example, Sackett, Holm, and Landesman-Dwyer (1975) have identified variables such as parity, age, medical treatments both before and during pregnancy, sex of fetus, and duration of pregnancy as factors contributing to abortion and stillbirth in a colony of M. nemestrina housed in a harem-breeding situation. Total fetal loss in the sample was 14.4% of total conceptions. This compares with data from the Wisconsin rhesus

TABLE 1.   NUTRITIONAL   DATA   (LABORATORY   CHOW)

Laboratory	Protein Content	Amount	Supplement
Hazleton	25%	0.33 lb/day	½ vitamin sandwich* + ½ apple
Oregon RPRC	15%	ad lib	none
Washington (Medical Lake)	15%	ad lib	none
Washington (Seattle)	25%	ad lib	none
Wisconsin RPRC	15%	ad lib	none

*Ingredients (750 g powdered milk, 44 g folic acid, 100 ml ascorbic acid and 1½ gal molasses) mixed with water into consistency of peanut butter and spread on bread.

colony of 15-21% (depending on the year of sample) without intervention and 5.2% with intervention by the veterinary staff (see Chapter 1). Combined fetal loss for a colony of rhesus and crabeating macaques and African green monkeys at Hazleton was 12.4% stillbirths and 5.8% abortions, with 81.8% live births for the years 1973-76. Five of 26 stillbirths in this colony were confirmed breech deliveries.

Most births for all species of monkeys occur during the night or early morning hours (as in the wild), a time when interruption and intervention are minimized. As part of a study to investigate the effects of prenatal stress (see Chapter 2), 49 M. nemestrina were shipped from the Medical Lake Field Station to the Infant Laboratory in Seattle at 125-130 days gestation and placed in single cages until parturition. All subjects were multiparous females and had been singly housed prior to conception. The females were observed every 30 min, 24 hr per day, via infra-red closed-circuit television for clinical signs of labor. When labor began, technicians were called in to intervene immediately after birth. The birth statistics are as follows: 40 live births with proper presentation = 81.6%; 1 stillbirth, not breech = 2%; 8 breech deliveries = 16.3% (3 were resuscitated, 5 died). The high percentage of breech deliveries in this sample suggests that it might be wise to institute a program similar to Dr. Mahoney's (Chapter 1) in an attempt to minimize this pregnancy outcome.

PERINATAL PERIOD

Birth Weight

A useful clinical tool in assessing neonatal "fitness" or ability to survive is birth weight. Birth weights of M. mulatta and M. fascicularis at Hazleton are summarized in Table 2, and birth weight distributions for M. mulatta at Wisconsin and M. nemestrina at Washington are presented in Figure 1. The Wisconsin rhesus macaques are remarkably similar in their distribution of birth weights to the Washington pigtail macaques, but the Hazleton rhesus neonates are considerably larger than those from the Wisconsin colony. There is no definite explanation for this discrepancy, but factors such as selective breeding and culling practices as well as age and parity of the females could influence birth weight. Another factor could be housing. At both Hazleton and Washington, females caged alone produce heavier infants than those housed in groups (Table 3). While the reason for this has not been documented, a reasonable theory might be that females housed in single cages are subjected to less stress and possibly exercise less than females housed in a group situation, the effect of which could be a longer gestation and, therefore, a heavier baby. The mean gestation for gang-housed pigtail macaques at the Medical Lake colony was 167 days (Sackett et al., 1975), while the mean gestation for 52 singly-caged females was 171.46 days (Dr. Richard Holm, unpublished data). Prolonged gestation could have important effects on the percentage of neonates to survive either when reared with the natural mother or separated from the mother in a nursery setting. For example, while all three laboratories reported overall survival in their nurseries of approximately 95%, all reported an estimated lower percentage of success in attempting to rear low-birth-weight or premature neonates. In the Infant Laboratory, survival rates for neonates in the lowest 10th centile of weight for the colony were 38% in 1974, 54% in 1975, and 86% in 1976. Increased survival rate over the years can be attributed to

TABLE 2.   BIRTH WEIGHTS AT HAZELTON, 1965-1976

Species	Sex	N	Mean (g)	Range (g)
M. mulatta	male	162	512.9	358-806
	female	151	474.9	287-678
M. fascicularis	male	123	356.3	161-546
	female	102	336.7	213-507

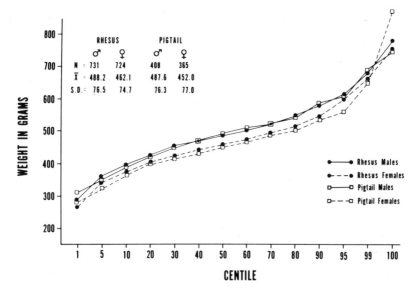

Figure 1.   Cumulative birth weight distribution for rhesus versus pigtail macaques.

increased intervention by the medical staff and to the fact that neonates in this "risk" category are of particular research interest in this laboratory and receive extraordinary attention.

Diet

The protocols for formula feeding of neonates varied among the four laboratories surveyed.   All use commercial formulas manufactured for human neonates but follow various feeding schedules based on staff availability and research emphasis.   Basic formula mixtures for normal animals 4 days of age and older are as follows:

Hazleton: 20 Kcal Similac/iron(R) (Ross Laboratory).

Oregon: 26 Kcal SMA(R) (Wyeth Laboratory).

Washington: 18.29 Kcal SMA. (This laboratory began their nursery rearing using the Wisconsin protocol with Similac. It was their impression that pigtail neonates use the SMA product more efficiently as indicated by stool firmness. This remains untested experimentally and is offered as information only.)

Wisconsin: 18.29 Kcal Similac/iron. (Diluted with 35 oz of water rather than 32 oz as product dictates to decrease the incidence of constipation.)

Prior to 4 days of age, three nurseries use either substitute formulas or diluted formulas for various lengths of time as follows:

Oregon: Premature infant formula (Meade-Johnson #291-02) given days 1-2.

Washington: Dextri-Maltose(R) (Meade-Johnson) constituted to 11.54 Kcal/oz every 2 hr for first 8 hr postpartum for neonates born in the laboratory and separated at birth. For neonates arriving from Medical Lake, the above formula every 2 hr for 3-4 feedings. (There is a rapid decline in blood-glucose levels if infants are not fed until noon of day 1.) Then equal parts SMA and Dextri-Maltose formula every 2 hr until 72 hr postpartum. Formula concentration equals 14.92 Kcal/oz.

TABLE 3.   MEAN BIRTH WEIGHTS IN GROUP VERSUS SINGLE HOUSING

Species	Sex	N	Mean birth weights
Gang housed			
M. mulatta	male	14	499
	female	14	461
M. fascicularis	male	27	350
	female	28	324
M. nemestrina	male	408	488
	female	365	452
Single-cage housed			
M. mulatta	male	58	521
	female	52	491
M. fascicularis	male	33	378
	female	21	361
M. nemestrina	male	30	541
	female	22	483

Wisconsin: 10% dextrose (anhydrous reagent A.C.S.) formula from
               noon of day 1 through 10 a.m. of day 2.   Formula
               concentration provides 11.54 Kcal/oz. Equal parts of
               Similac formula and 10% dextrose formula through
               day 3. Formula concentration provides 14.92 Kcal/oz.

To make comparisons between facilities, data on formula
consumption were calculated for kilocalorie intake and fluid intake
for the first 2 weeks of life.   Available data allowed a fairly
extensive examination of Hazleton and Wisconsin rhesus colonies
and is displayed as average intake.   The information from the
Oregon nursery is exhibited as maximum intake (from information
supplied) only for comparison.

The Hazleton and Wisconsin feeding regimens offer a liberal
amount of formula at each feeding, ad lib at Wisconsin and up to 25
ml at Hazleton, while Oregon restricts intake more severely
(Figures 2 and 3).

Weight gains at Wisconsin and Hazleton are compared in
Figure 4.   As noted in Figure 1, rhesus monkeys at Hazleton are
appreciably heavier at birth than those at Wisconsin.   Hazleton
males average 512.9 g (n=162) and females average 474.9 g (n=151),
while Wisconsin males average 488.2 g (n=731) and females average
462.1 g (n=724).   Over the 2-week period both sexes at Hazleton
gain an average of 20% over their birth weights, while at Wisconsin

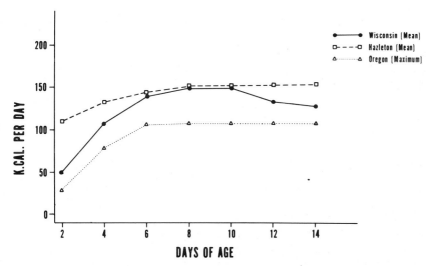

Figure 2.   Estimated kilocalorie ingestion for M. mulatta.

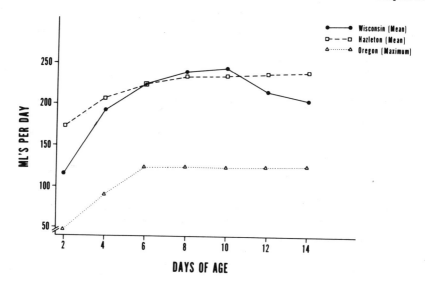

Figure 3.   Estimated water (formula) consumption for M. mulatta.

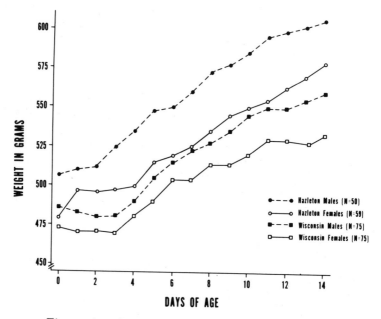

Figure 4.   Average weight gain for M. mulatta.

males gain an average of 15% and females, 12.5% over birth weights.   In addition, Hazleton neonates show a weight gain beginning on the first day of life while those at Wisconsin take an average of 4 days to regain and surpass birth weight.   It has been estimated that a 500-g rhesus macaque requires 40 Kcal/day for maintenance of basal metabolism and 200 Kcal/kg for adequate growth to occur (Bourne, 1975).   In addition, van Wagenen and Catchpole (1956) reported an average weight gain of 47.7 g over 2

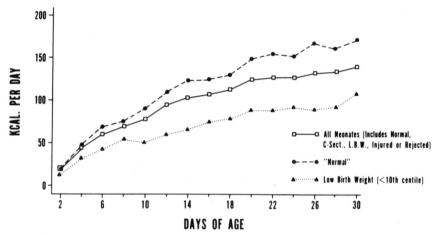

Figure 5.   Kilocalorie ingestion for male M. nemestrina.

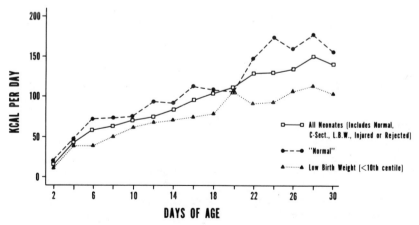

Figure 6.   Kilocalorie ingestion for female M. nemestrina.

weeks for a sample of 34 rhesus neonates. The Hazleton animals exceed that average (50 g for males, 49 g for females), while the Wisconsin animals fall below it (39 g for males, 30 g for females). However, no comparative anatomical or physiological data are available from these facilities, nor is there evidence that rapid weight gain in itself is positively correlated with "fitness." Indeed, the high survival statistics from these colonies (Hazleton = 95.0% for all rhesus, 98.7% excluding experimental subjects (1965-1976); Wisconsin = 95.6% for 30 days if alive after first day of life) indicate that good results can be obtained despite variability in diet protocols.

The kilocalorie ingestion data for a sample of M. nemestrina reared for 30 days in the Infant Primate Research Laboratory are presented in Figures 5 and 6, and fluid intake data for the same sample are shown in Figures 7 and 8. Low-birth-weight infants ingest less formula per day over the entire 30-day period than do other neonates in the nursery. Factors influencing growth of these subjects are discussed at length in Chapter 14. It should be noted that food intake on day 1 for these animals is not included in the figures because many animals are shipped to the Infant Laboratory from Medical Lake the first or second day of life, reducing sample size and increasing error probability. Clearly, however, mean intake values for the M. nemestrina in the Washington colony are appreciably lower during weeks 1-2 than those in either the Hazleton or Wisconsin rhesus samples.

There were no comparative weight gain data for M. fascicularis. The Infant Laboratory rears small numbers of this species each year, but insufficient numbers make interpretation of data difficult. However, weight gain and formula consumption data for a sample of 44 male and 35 female M. fascicularis from birth through 2 weeks of age at Hazleton Laboratories are presented in Figures 9 and 10. It should be noted that mean birth weights for both sexes in this sample are appreciably higher than overall average birth weights for this species at Hazleton (males = 356.3 g, n=123; females = 336.7 g, n = 102).

METHODS OF FEEDING

The following information from the Washington nursery may be useful to new nursery facilities. The other nurseries surveyed here use similar procedures.

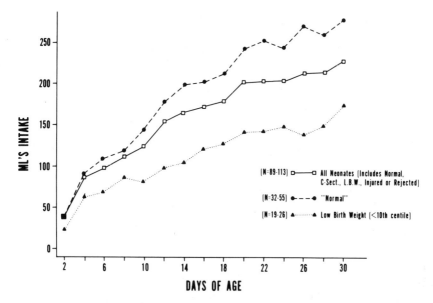

Figure 7.   Water (formula) consumption for male M. nemestrina.

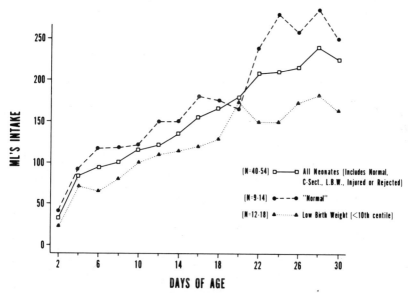

Figure 8.   Water (formula) consumption for female M. nemestrina.

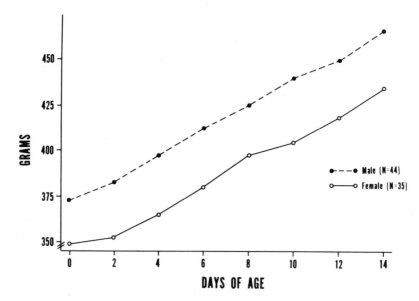

Figure 9.   Average weight gain for M̲. fascicularis.

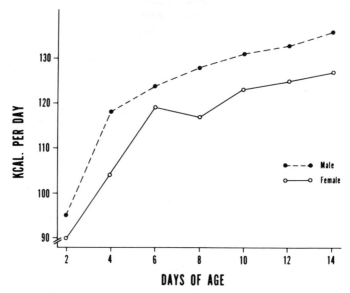

Figure 10. Kilocalorie consumption for M̲. fascicularis.

Bottle Feeding

Infants are bottle-fed using a small 2-oz bottle manufactured for sale in pet supply houses (Poly-Nurser Products Corporation, Brooklyn, NY). A hole is put in the nipple by heating a 22-gauge syringe needle until it glows, then punching it into the end of the nipple. A 24- to 27-gauge needle is used to make the hole for feeding premature and low-birth-weight animals. The bottle and nipple can be boiled but not autoclaved for sterilization.

Gavage Feeding

Neonates less than 150 days of gestation or less than the tenth centile in birth weight are considered to be at extreme risk for aspiration pneumonia due to poor coordination of sucking and swallowing; therefore, a pediatric 3.5-5.0 French infant-feeding tube is inserted by either the nasal or the oral route. The residual fluid in the stomach is then measured and replaced by drawing the fluid into the syringe attached to the tube. Feeding is begun with 3-5 ml of fluid, increasing the amount by 1- to 2-ml increments every 2 hr as tolerated by the animal indicated by the residual amount of fluid measured in the stomach. Gavage feeding is often necessary for as long as 7-10 days. When attempting initial bottle feeding following gavage feeding, use small amounts of fluid (3-5 ml) and follow the progress of the neonate carefully.

Position of the Infant

The infant is held in an upright position to aid in swallowing. Wrapping the infant in a diaper for feeding or having a diaper on the ventrum helps since the grasping-clasping-righting reflexes are developed by 140 days of gestation or earlier. Most infants can be trained to self-feed from bottles by 7-14 days of age. However, premature animals occasionally take up to 45 days before motor control is sufficiently developed to make self-feeding possible (unpublished observation).

APNEA AND OXYGEN THERAPY

Occasionally, an acute incidence of apnea occurs in premature neonates. (For aid in monitoring apnea, refer to Chapters 15 and 16.) Oxygen therapy should be used following vomiting and aspiration or when the neonate exhibits signs of Idiopathic Respiratory Distress Syndrome, seen in human neonates. Oxygen therapy is initiated at 50-70% oxygen, determined by relief to the individual animal. In severe cases, 100% oxygen levels have been used for as long as 24 hr with no incidence of retrolental fibroplasia having been observed in our nursery. High oxygen levels are achieved by one of two methods: 1) placing the newborn under a plastic dome in the isolette and administering heated and humidified oxygen, or 2) using a respirator (Harvard Apparatus Rodent Respirator Model 680) with a 3.5-mm esophageal tube (Cole tube) at 60-90 strokes/min with 2-3 cc/stroke. Care must be taken with this second procedure since tidal volume in a 300-g premature infant is 1-1.5 cc on day 1 (personal communication from Dr. Robert Guthrie). Overinflation can lead to capillary hemmorhage. Settings vary depending on the tubing length and compliance, so calibration must be done carefully. This procedure has been used with moderate success.

DIET ADDITIONS, GROWTH, AND WEANING

Wisconsin and Oregon nurseries add commercial liquid multivitamins (formulated for human neonates) at a rate of 3-4 drops daily until the infant is 90 days of age. The Hazleton and Washington facilities currently add no vitamins to their formula.

Protocols for introduction to solid foods and weaning vary greatly among the facilities (Table 4).

Weight gains for the Hazleton and Wisconsin rhesus monkeys are compared in Figure 11. Both males and females at Hazleton show a nearly linear weight increase across the entire first 6 months of life. While animals from the Wisconsin colony show a slightly less linear weight increase over the 6-month period, the differences in linear growth are minimal. However, it is of interest to note that average weights for females from the Hazleton colony exceed weights for males from Wisconsin. Factors other than food consumption that might account for these discrepancies will be discussed below.

Data on weight gain by sex were unavailable from the Oregon nursery facility; however, from information on average weight gain from 100 nursery-reared rhesus macaques of unidentified sex (Fuquat, Hall, and Jones, unpublished data), we extrapolated estimated weights for a 6-month period and compared them with combined-sex weight data from Hazelton and Wisconsin. The results are shown in Figure 12. Animals from the Oregon colony sample do not show a near-linear growth pattern before about 16 weeks of age. The reason is unclear, but one factor could be that

TABLE 4..   WEANING AND INTRODUCTION TO SOLID FOODS

Hazleton:

Solids	Days 16-30:	small amounts of vitamin sandwich (see above), fruit and monkey chow.
	Days 31-120:	¼ vitamin sandwich, ¼ piece fruit, 4-10 biscuits of monkey chow.
	Day 121 on:	½ sandwich, ½ piece fruit, 10-24 biscuits of monkey chow.
Formula	Day 46 on:	change from Similac formula to homogenized milk.
	Days 46-90:	100 ml milk offered TID*.
	Day 91-180:	ml milk offered BID.
	Days 181-365:	100 ml milk offered SID.

Oregon:

Solids	Day 50 on:	introduction to fruit and monkey chow (no data on amount fed).
Formula	Day 51-260:	offered 50 ml formula TID; infant cereal added to formula at 90 days.
	Day 260:	weaned off formula.

Washington:

Solids	Day 21 on:	monkey chow ad lib and fruit SID.
Formula	Self-feeding to 90-135 days: 150 ml offered TID.	
	Day 90-135 for 2-4 weeks: commence weaning by diluting formula to 60% SMA/ 40% water, then 40% SMA/60% water; time of weaning dependent on individual animal, solid food consumption, and weight gain.	

Wisconsin:

Solids	Day 46 on:	introduction to monkey chow ad lib.
Formula	Days 31-180:	offered 170 ml BID.
	Day 180:	weaned off formula.

*TID = three times per day; BID = twice per day; SID = once per day

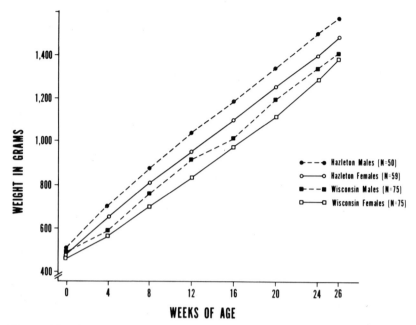

Figure 11. Six-month weight gain for male versus female nursery-reared M. mulatta.

the availability of food is more restricted at Oregon (Figures 2 and 3).

Survival statistics and other epidemiological information were not available from the Oregon nursery; therefore, it cannot be determined what impact the lower weight gain has on this particular colony of rhesus macaques.

Data concerning weight gain for pigtail monkeys in the Washington nursery are examined closely in Chapter 14. Summary statistics from that sample indicate that mean weight gain through the first 16 weeks of life (until weaning) totals 650 g for normal males and 509 g for normal females. During that same period, low-birth-weight males gain an average of 558 g and low-birth-weight females gain 541 g. A comparison of low-birth-weight versus normal animals was not available from Hazleton and Wisconsin. However, overall estimated weight gain for the rhesus colony at Hazleton totaled 675 g for males and 625 g for females for the first 16 weeks of life. Weight gains for Wisconsin rhesus males and females totaled approximately 525 g for the same period. Weight gain for the rhesus colony in the Oregon nursery was estimated at 400–425 g from available data for the 16–week period.

The weight gain averages for the sample of 44 male and 35 female M. fascicularis (Figure 9) from the Hazleton Laboratories exhibit the same near-linearity across the first 6 months of life as do that facility's weight gains for rhesus macaques. Both sexes double their birth weight by about 10 weeks of age and weigh an average of 1,190 g (males) and 1,120 g (females) at 6 months of age.

REARING ENVIRONMENT

Isolettes

At Oregon, Washington, and Wisconsin, newborn macaques are placed in isolette units manufactured for human newborns. In the Oregon nursery, all neonates live in isolettes for the first 2

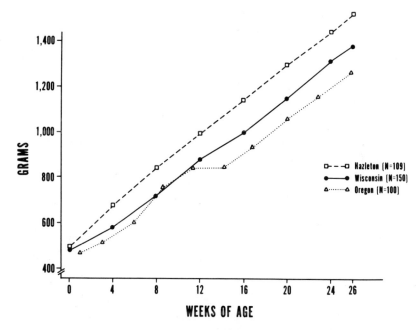

Figure 12. Estimated weight gain for M. mulatta in three nurseries.

weeks of life; temperature is 85-90°F. In the Washington nursery, all neonates live in isolettes with temperature set at 92-94°F and humidity at 60-80%. After the animal's rectal temperature stabilizes at 98.5°, isolette temperature is reduced in 2-3° increments. (Refer to Chapter 10 for detailed examination of physiological parameters.) The infant is removed from the isolette when isolette temperature equals 86°F and the animal's temperature remains normal. At Wisconsin, all neonates under 350 g at birth are placed in isolettes, and other infants only as required. Temperature is 85-87°F and humidity is 60-90%. Hazleton Laboratory places newborns in a clear plastic isolator of their own design with a heating pad set on low heat for 15 days. External room temperature is 76-78°F. Cage size is 1.13 cu ft.

Room Temperatures

Following initial isolette rearing, room temperatures are maintained as follows: at Hazleton, 76-78°F (estimated); at Wisconsin, 74°F. At Washington, the nursery room is set at 82°F, and heating pads are kept on low heat until diurnal rectal temperatures are stable. The rearing room is set at 78°F. (It had been maintained at 72-74°F until 1972, but incidence of unspecified chronic upper respiratory infections dictated increase in environmental temperature, which alleviated the problem.)

Cage Sizes for Normal Rearing

At Hazleton, rearing cages measure 1.88 cu ft; at Oregon, 4.35 cu ft; at Washington, 2.43 cu ft to 60 days, then 4.99 cu ft or 12.7 cu ft depending on experimental demands; and at Wisconsin, 2.60 cu ft up to 30 days, and then 6.00 cu ft.

Light-Dark Cycles

Hazleton has 12 hr light/12 hr dark; Oregon has no set cyclicity; Washington has 14 hr light/10 hr dark; and Wisconsin has 16 hr light/8 hr dark.

Diapers

All facilities surveyed place diapers in cages but for varying lengths of time. Use of diapers is, in all probability, a wise practice both as an aid in maintaining temperature and as a means of avoiding urine scalding, which plagues neonates. Diapers are removed permanently at 3-4 months at Hazleton, 21 days at Oregon, 4 months at Washington, and 15 days at Wisconsin.

Socialization

The need for peer contact for developing species-typical behaviors in macaques has been well documented by several investigators. Chapters 19 and 20 deal with this issue in some detail.

COMMON PROBLEMS IN NURSERY FACILITIES

Diseases

The most common nonrespiratory health problem enountered by the four laboratories surveyed is diarrhea. This problem stems from many causes and is extremely difficult to control. Iatrogenic diarrheas are often caused by destruction of normal gut flora. Infectious disease problems are numerous and most commonly bacterial in origin. Viral and parasitic infections are encountered less frequently but not uncommonly. To compound the problem, stress (environmental, social, experimental) often is a precursor to infectious disease because it lowers resistance. Another common diarrhea in monkeys results from dietary changes or overconsumption. It can usually be corrected by limiting available food, or diluting formula as required. Metabolic diseases are uncommon but do occur in monkeys. Hazleton Laboratories report two incidences of lactose intolerance in rhesus neonates, relieved by substituting soy-base human formula (Dr. Dan Dalgard, unpublished data).

The supportive therapies suggested below are simply guidelines, and are by no means comprehensive, exhaustive, or detailed. Assessment of individual animals must be made by qualified veterinary staff whenever possible.

Fluid Therapy

Intravenous or subcutaneous fluid support is indicated for acute dehydration when an animal is severely depressed or comatose. Infusion of various solutions (i.e., Ringers, lactated Ringers, water with dextrose as determined by blood chemistry analysis) are commonly used fluid therapies when oral administration is deemed disadvantageous or impossible. Because this is not a clinical paper dealing with the complexities of the problem and because individual interpretation varies, the reader must rely on other sources for detailed recommendations.

The four nurseries surveyed use various foods and products when oral therapy is indicated for moderate dehydration. Citrus fruits and/or apples are commonly used and usually are readily eaten. Hazleton recommends Tang(R) as an appetite stimulant. The Infant Laboratory commonly uses oral electrolyte solution (Pedialyte(R), Ross Laboratories) when indicated, either as supplied or as a dilutant for formula in 50:50 mixture. Animals seem to accept this product readily and assimilate it well. Cultured milk products (kefir, yogurt, milk) with active lactobacillus have been used successfully, either as supplied or diluted with water or oral electrolyte solution.

Heat Therapy

A common problem associated with dehydration caused by diarrhea is hypothermia. Use of a heating pad on low heat often relieves the problem if the animal is not depressed. For acute dehydration-depression, an isolette or other warm-humidified chamber is useful. Using a heating pad when animals are immobile can lead to shock from overheating the ventrum with concomitant shunting of blood to the ventral surface.

SUMMARY

The differences in rearing environments among the four nursery facilities confound the issue of diet manipulation and make interpretation of weight gain information difficult. It seems logical, for example, to assume that ambient room temperature differences could influence weight gain via heat radiation from the animal with commensurate calorie consumption increase or decrease. Partial insulation with a diaper could also influence energy requirements. Macaques have been shown to exhibit higher activity during the daytime versus night hours. Longer days with shorter nights, allowing the animals increased opportunity for high activity, could also influence energy requirements. In short, environment must be taken into account when comparing data because of the potential for influencing results in a subtle or unknown fashion.

I would hope that this survey will aid and prompt others to evaluate and share their own data in a quantitative and objective fashion, for it is through such a process that we promote the scientific knowledge essential for successfully rearing nonhuman primates. Only through shared information can the goal of providing optimal care be approached and the science of nursery rearing be advanced to the mutual benefit of all.

ACKNOWLEDGMENTS

This survey would not have been possible without the cooperation and consideration of the following individuals and laboratories: Ms. Jeannette Reeves and Dr. Dan W. Dalgard of Hazleton Laboratories America, Inc.; Mr. Frank Fuquay, Dr. Arthur Hall, and Mr. Richard Jones of the Oregon Regional Primate Research Center; Mr. Chris Ripp, Jr. of the University of Wisconsin Department of Psychology Primate Laboratory; and Dr. C. J. Mahoney (now at the Laboratory for Experimental Medicine and Surgery in Primates, NY) of the Wisconsin Regional Primate Research Center.

I would also like to express my thanks to the entire staff of the Infant Primate Research Laboratory at the University of Washington for their assistance in the preparation of this manuscript and especially for their industrious and professional approach to the care of the neonates in our facility.

This research was supported by NIH grants HD08633 and HD02274 from NICHD Mental Retardation Branch and RR00166 from the Animal Resources Branch.

REFERENCES

Bourne, G. H. Nutrition of the rhesus monkey. Pp. 98-114 in: The Rhesus Monkey, Vol. II., G. H. Bourne (Ed.), New York: Academic Press, 1975.

Sackett, G., Holm, R. and Landesman-Dwyer, S. Vulnerability for abnormal development: Pregnancy outcomes and sex differences in macaque monkeys. Pp. 59-76 in: Aberrant Development of Infancy: Human and Animal Studies, N. R. Ellis (Ed.), New York: Halsted Press, 1975.

van Wagenen, G. and Catchpole, H. R. Physical growth of the rhesus monkey (Macaca mulatta). Amer. J. phys. Anthrop. 14:245-273, 1956.

CHAPTER 14

PONDERAL GROWTH IN COLONY- AND NURSERY-REARED PIGTAIL MACAQUES (Macaca nemestrina)

G. P. Sackett, R. A. Holm, and C. E. Fahrenbruch

Regional Primate Research Center and
Child Development and Mental Retardation Center
University of Washington
Seattle, Washington

Computerized weight records are available for about 1,500 mother-reared pigtail macaques (Macaca nemestrina) born at the Primate Field Station located at Medical Lake, Washington. In addition, weights are available on over 150 hand-reared monkeys who were separated from their mothers at birth and housed in the Infant Primate Research Laboratory nursery. This data bank is a major source of normative growth information. It also may play an important role in general health care practices with infant and older animals.

During the past 6 years, we have studied this weight information along with other data concerning health and reproductive outcomes. The purposes of this work include 1) screening the breeding colony for dams and sires likely to produce low-birth-weight or premature offspring (Sackett and Holm, 1977); 2) providing a developmental index to measure the effects of parental characteristics, prenatal factors, and varied rearing conditions on growth; and 3) developing a computer health monitoring system to predict illness from changes in weight of each monkey in the colony. The purpose of this chapter is to present our ponderal growth norms for pigtail monkeys, and to show how growth varies with housing conditions and status at birth. We believe that knowledge about birth condition and environmental effects on growth are essential for at least two reasons. First, typical (normative) values averaged over a total primate colony will reveal a poor picture of reality if different rearing conditions have large effects on the normative measure. Second, the accuracy of

187

statistics for identifying individuals whose growth pattern or current weight value is abnormal will depend on identifying important variables which influence growth rate and/or adult values.

HOUSING CONDITIONS

The Medical Lake breeding facility and husbandry practices have been described in detail by Blakley, Morton, and Smith (1972). Breeders are housed in large "harem" rooms containing six to 12 females and one male. Infants are born in these rooms and live with their mothers until weaning at 3-8 months of age. They are then housed in rooms containing a number of agemates until assigned to the breeding program or to an experimental project. Although some females with timed pregnancies have been maintained in single cages, their offspring constitute only a few percent of total births and thus have a negligible effect on the data presented here.

The Infant Laboratory nursery contains three general types of infants. 1) Normal nursery infants are naturally delivered and are removed from their mothers immediately at birth. Most of these newborns are healthy, normal, full-term deliveries. Other healthy newborns are naturally delivered at Medical Lake and come to the nursery within 1-3 days of birth. Most of these animals serve as subjects in nonintrusive behavioral development studies. 2) Low-birth-weight infants also come from natural births at Medical Lake, but are either premature, low in birth weight, injured, or sick. Such high-risk infants almost always die if left with their mothers. Upon detection, they come to the nursery and receive intensive care as required by their particular clinical problem. Growth data for these infants will be presented in this chapter. 3) Caesarean section (C-section) infants often have mothers that have been experimentally manipulated, or are themselves subjected to treatments after delivery. Some weight data for full-term, normal-birth-weight, C-section infants will be presented in this chapter.

Our nursery practices have been described in Chapter 13. Briefly, infants are hand-fed until they are about 14-21 days old, at which time they can feed themselves. After attaining full self-feeding abilities, infants are removed from their special nursery cages and housed individually or in pairs in general colony cages. Sick neonates live in environment-controlled isolettes. Most animals leave this situation by 2-4 weeks after birth. By 30 days of age almost all infants in the nursery begin a standard battery of developmental tests for growth and behavior. This includes 4-5 days per week of social interaction in a large playroom. Developmental testing lasts until the infants are 9–12 months old.

SUBJECTS AND MEASUREMENT PROCEDURES

The number of individuals and the summary data on their weights are presented in Table 1. Birth weight measures are based only on animals surviving the first day of life, or healthy animals assigned to experiments at birth. Developmental weights for breeding colony subjects were collected at 2- to 3-month intervals in association with tuberculosis testing and at weaning. Thus, data for colony subjects are mixed longitudinal–cross sectional values in which all individuals contribute more than one weight but not all are measured at the same ages. Weights for nursery subjects were taken daily during the first month of life, 2–3 times per week during months 2 and 3, and once per week thereafter. As almost all nursery subjects contribute some data for each week of age in the data presentation, the nursery data are probably more accurate in estimating true norms than data for breeding colony animals. However, inter-subject variability did not differ markedly (nor with statistical significance at the $p = 0.05$ level) in any comparison

TABLE 1.   SAMPLE SIZES FOR MALE AND FEMALE M. nemestrina BORN IN THE COLONY OR IN EACH OF THREE NURSERY GROUPS, YIELDING THE 16-WEEK WEIGHT DATA.

| Measure | Sex | Breeding Colony | Nursery | | |
			Normal	Low Birth Weight	C-Section
Number of Individuals	Male	337	62	14	12
	Female	319	44	16	11
Number of Weights	Male	1254	2641	1530	988
	Female	1193	2226	1681	1011
Average Weights per Individual	Male	3.7	42.6	109.3	82.3
	Female	3.7	50.6	105.1	91.9

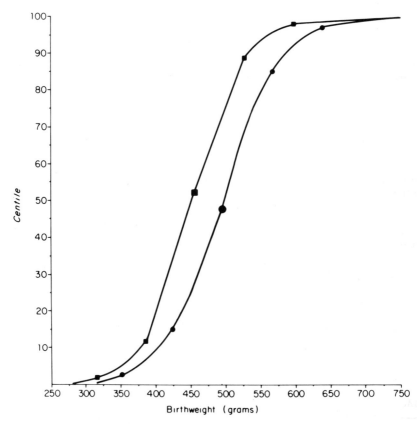

Figure 1.   Cumulative birth weight distributions for male (circles,
             N = 337 weights) and female (squares, N = 319 weights)
             M. nemestrina.   Large symbols show means, small
             symbols show 1 and 2 standard deviations from the
             mean.

weeks.   Thus, comparisons between groups probably are not
affected in any major way by the difference in data collection
schedule.

Weights entering into growth curves were selected by two
criteria.   First, the animal must not have been under medical
treatment within 7 days of the weight measurement. Second, the
animal must not have died from any type of nontraumatic natural
cause within 14 days of the measurement.   These criteria were
chosen to maximize the chance that our growth norms would

reflect a valid picture for healthy animals. An exception to the first criterion was made for low-birth-weight and C-section new-borns. As these animals were all under some type of medical treatment during the first 1-2 weeks of life, data for these ages were included provided that the animal did not die of natural causes within 14 days of the weight determination.

BIRTH WEIGHT

The cumulative distributions of birth weights for pigtail macaques in the breeding colony are shown in Figure 1. Males were 8% heavier (mean = 490 g, S.D. = 67) than females (mean = 450 g, S.D. = 70) at birth. The female cumulative curve made an excellent fit to a normal distribution, while that for males was skewed with an excess of low birth weights (p 0.025 by chi-squre goodness-of-fit test). In fact, the lowest birth weights were for males and the highest were for females. The birth weight cutoff points for the lowest 10 centiles were 405 g for males and 380 g for females. These values are used in this chapter to define low-birth-weight newborns.

GROWTH OF COLONY-BORN ANIMALS

Weight data from birth through 72 months of age were available for 450 males (3610 values, mean = 8.0 values/animal) and 410 females (3395 values, mean = 8.3 values/animal). A regression analysis was applied to the values within each sex, yielding significant linear, quadratic, and cubic components. Mean weight per month, standard error of the mean, and the growth curve predicted from the regression analysis are shown in Figure 2. Male weights exceed female weights during the 72-month period, with differences becoming especially marked after 2 years of age. At about 3.5-4 years of age males showed an abrupt rate change, illustrating a marked "adolescent" growth period. Females showed a slight acceleration in growth rate before 3.5 years but did not show any changes comparable to those of males.

Figure 3 presents mean weight values through 48 months of age as a function of birth weight centiles. Sample sizes for the lower-birth-weight groups were too small to continue the curves beyond 4 years of age. Both males and females in the lowest 10 centiles of birth weight remained lower in weight than any other birth-weight group throughout the age range studied. Further, both low-birth-weight groups had the lowest rate of weight gain across this 4-year period. Thus, low-birth-weight monkeys reared with mothers under the Medical Lake conditions failed to exhibit any evidence for a "catch-up" toward average body weight shown by all but the lowest-birth-weight humans (Tanner, 1963).

Males in the heaviest 10 centiles for birth weight remained the heaviest animals throughout this 4-year period. They also gained weight faster than any other male birth weight centile

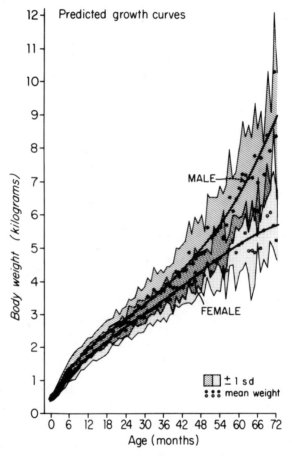

Figure 2.   Six-year growth curves showing mean monthly weights, and weights predicted by linear, quadratic, and cubic regression for colony-reared M. nemestrina. (Shaded areas show standard error of the mean.)

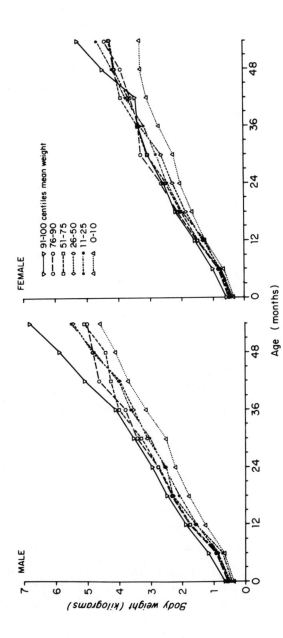

Figure 3. Four-year growth curves for breeding colony M. nemestrina that were at different centile positions in the birth weight distribution.

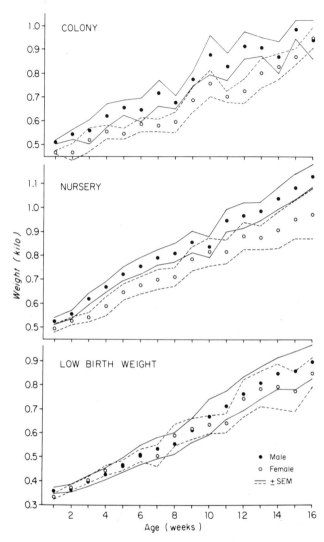

Figure 4.   Growth curves during early infancy for males and
females reared by mothers in the breeding colony
compared with normal and low-birth-weight nursery-
raised monkeys.

group, and exhibited an adolescent growth spurt about 6 months before the other male groups. Thus, for males heavy birth weight was related to faster growth, while low birth weight produced retarded growth up to the end of adolescence. For females heavy birth weight did not result in accelerated growth, although low-birth-weight females were retarded in weight gains.

These data show that different growth norms are required to adequately characterize the growth of colony animals that differ in birth condition. Furthermore, the fact that sex differences appeared in both absolute weight and the form of the age-weight curve suggest that weight norms must be maintained separately for males and females--a practice that is not always done with macaque growth data.

GROWTH OF BREEDING COLONY VERSUS NURSERY INFANTS

Growth during the first 16 weeks of life was compared for mother-reared colony and hand-reared nursery infants (see Table 1 for sample sizes). Although nursery birth weights were slightly higher, none of the differences were statistically reliable (males: colony = 490 g, S.D. = 67; nursery = 505, S.D. = 52; females: colony = 450, S.D. = 70; nursery = 460, S.D. = 55). Figure 4 shows that nursery males and females were heavier than their colony counter-parts during almost all of the first 16 weeks of life. Unfortunately, sample sizes for colony weights were too small at any particular week to allow comparisons beyond week 16. However, average weight at 11-12 months of age showed that nursery monkeys were still heavier than those raised in the colony: colony males weighed 1.75 kg compared with 2.20 kg for nursery males, while colony females weighed 1.55 kg compared with 1.85 kg for nursery females.

These data show that growth norms during infancy varied markedly with mothering/housing conditions. This may be due to diet, activity level, social group composition, or other factors which merit study in their own right. However, it seems clear that regardless of the reasons for these differences, norms based on one group will not yield an accurate picture of ponderal growth velocity values (grams weight gain per week) shown in Table 2. Normal nursery males gained weight much faster than colony males, with little overall difference between females. Nursery animals had relatively uniform weekly weight gains, but colony infants showed extreme deviations in weekly gain during the first 4 months of life.

Growth curves comparing males and females during the first 16 weeks of life are shown in Figure 5. Males were heavier than females in almost all weeks for both colony and nursery infants, with the sex difference magnitude almost identical. Averaged over all 16 weeks, colony males were 87.4 g heavier than females, while nursery males were 86.6 g heavier than females. Another comparison of interest showed that colony males averaged only 7.3 g

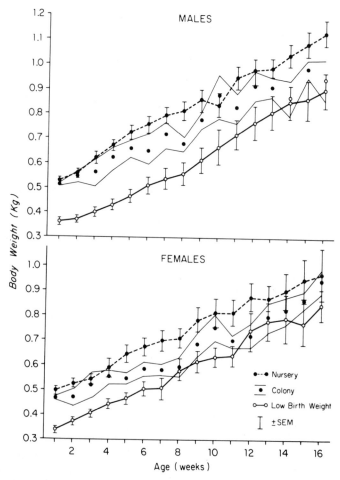

Figure 5.   Sex differences in growth during early infancy for breeding colony, normal nursery, and low-birth-weight monkeys.

TABLE 2.   WEIGHT GAIN PER WEEK (AVERAGE GRAMS) FOR
            COLONY AND NURSERY-RAISED PIGTAIL MON-
            KEYS.

	MALES			FEMALES		
			Low			Low
		Normal	Birth		Normal	Birth
Week	Colony	Nursery	Weight	Colony	Nursery	Weight
1-4	36.7	47.7	23.0	28.3	31.0	36.3
5-8	14.0	35.2	31.8	10.0	30.5	36.5
9-12	59.0	39.2	52.0	32.5	42.8	36.0
13-16	6.8	40.2	33.0	55.5	22.8	26.5
Overall Average	29.1	40.6	34.9	31.6	31.8	33.8

heavier than nursery females during the first 4 months of life.
However, nursery males averaged 79.5 g heavier than colony males,
while nursery females averaged 79.2 g heavier per week than
colony females.

Taking these group and sex differences together, we can
conclude that nursery-rearing produces heavier infants than
colony-rearing regardless of sex, and males are heavier than
females within each rearing group. The high weight for nursery
animals is illustrated by the finding that nursery females were
almost as heavy as colony males during the early part of infancy.

Unfortunately, data are not yet available to determine
whether nursery-rearing has permanent effects on body weight.
Thus, two important questions to be answered in the future concern
the age of the adolescent growth spurt for colony versus nursery
males, and adult weights of colony- versus nursery-reared infants.

BIRTH CONDITION EFFECTS IN NURSERY INFANTS

Low Birth Weight

Figure 4 shows growth curves for low-birth-weight compared
with normal nursery males and females. Low-birth-weight males

remain lighter than normal males throughout the first 16 weeks, while low-birth-weight females appeared to be catching up with their normal counterparts by the end of the first 4 months of life. This can also be seen in the velocity data shown in Table 2. Normal nursery males exhibit faster weight gain during most of the 16-week period, while low-birth-weight females gain weight faster than normal females in most of the time periods.

Figure 5 shows sex differences between normal and low-birth-weight infants. As noted in the discussion above, normal nursery males average 87.4 g per week heavier than females. However, low-birth-weight males average only 20.2 g heavier per week than low-birth-weight females, and the male curve does not differ reliably from that for low-birth-weight females in any of the first 16 weeks of life.

These results show that even with special nursery health care procedures, animals with a very low birth weight remain low in weight during the first part of infancy. However, these results differ in one important way from those presented for mother-reared colony animals in Figure 3. Under colony conditions neither males nor females exhibited evidence for a catch-up toward average weights over the first 4 years of life. Under nursery conditions males also did not show any early evidence of weight catch-up, but by 16 weeks low-birth-weight females were within the range of average weight for normal nursery females. Thus, it appears possible that special health care benefits pigtail females, but not males, in a high-risk birth condition.

CAESAREAN-SECTION DELIVERY

Figure 6 presents 16-week growth curves for normal nursery infants and infants delivered at full term by C-section. These C-section infants were healthy at birth and were not subjected to any special interventive procedures other than isolette-rearing for the first 1-2 weeks of life. C-section males weighed less than normal males in almost all weeks, and showed a markedly lower rate of gain in weeks 11-16. C-section females also weighed less than normals, but seemed to be catching up to normals by week 16. These data suggest that even for healthy newborns, C-section delivery may slow the growth rate during early infancy. The results are also similar to those for low-birth-weight animals in that males seem to be affected to a greater extent than females. Again, even with special nursery care, different norms may be required to assess the progress of growth in animals that vary in birth condition and in animals of different sex.

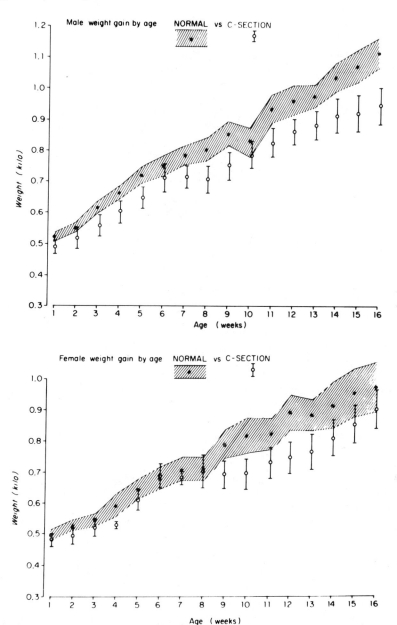

Figure 6.   Growth curves during early infancy for normal nursery males and females compared with nursery-raised infants delivered at full term by Caesarean section.

CONCLUSIONS

Although heavy body weight per se is not necessarily a desirable biological goal, it seems clear that in primate colonies small macaques are at a disadvantage. In humans, low birth weight can indicate a number of physical and behavioral abnormalities (e.g., Illingworth, 1970). Low-birth-weight monkeys are similarly at high risk for death, disease, and developmental abnormalities (Sackett, Holm, Davis and Fahrenbruch, 1974). Small monkeys, even as infants, tend to be socially subordinate and poor competitors among their peers in attaining the physical resources of their environment. Finally, abrupt changes in body weight of infant or adult monkeys provide one of the few obvious signs of health or stress problems in individual monkeys living in large colonies. Thus, weight at birth, housing differences and other factors producing large deviations in absolute weight and growth patterns are an important source of information in primate colonies for both husbandry and scientific purposes.

In this chapter we have shown that housing conditions, perinatal risk factors, and sex differences interact in complex ways to produce variations in growth patterns. Scientifically, this information is important for understanding both biological and behavioral factors influencing the development of monkeys. In terms of primate colony mangement these results suggest that a simplistic overall pooling of data to obtain normative values about developmental and health status is inadequate. Rather, understanding the reasons for success of reproductive and infant-rearing programs depends on identifying the variables that produce significant effects on major health and developmental indices such as ponderal growth. These variables can then be included in decision and evaluation procedures involving prediction of disease risk, appropriate assignment of animals to research projects, and success in rearing and maintaining both low-risk and high-risk individuals.

ACKNOWLEDGMENTS

This research was supported by NIH grants HD08633 and HD02274 from NICHD Mental Retardation Branch and RR00166 from the Animal Resources Branch.

REFERENCES

Blakley, G. A., Morton, W. R., and Smith, O. A. Husbandry and breeding of Macaca nemestrina. Pp. 61-72 in: Medical Primatology, Vol. I, E. I. Goldsmith and J. Moor-Jankowski (Eds.), Basel: Karger, 1972.

Illingworth, R. S. Low birth weight and subsequent development. Pediatrics 45:335-339, 1970.

Sackett, G. P. and Holm, R. A. Effects of parental characteristics and prenatal factors on pregnancy outcome and offspring development of pigtail macaques. In: Maternal Influences and Early Behavior, R. W. Bell and W. P. Smotherman (Eds.), Jamaica, N. Y.: Spectrum Publications, Inc., 1977.

Sackett, G. P., Holm, R. A., Davis, A. E., and Fahrenbruch, C. E. Prematurity and low birth weight in pigtail macaques: Incidence, prediction, and effects on infant development. Pp. 189-205 in: Proceedings of the 5th Congress of the International Primatology Society, Tokyo: Japan Science Press, 1974.

Tanner, J. M. The regulation of growth hormone. Child Develop. 34:817-847, 1963.

CHAPTER 15

MONITORING AND APNEA ALARM FOR INFANT PRIMATES: APPARATUS

F. A. Spelman and C. W. Kindt

Regional Primate Research Center
University of Washington
Seattle, Washington

INTRODUCTION

A recent survey (Health Devices, 1974) listed the criteria for the acceptance of an apnea monitor for use with human patients. The monitor must detect the cessation of breathing reliably and accurately; it must produce an audible and visible alarm; it must be safe; and it should be sufficiently small to be unobtrusive in a hospital setting. Franks and colleagues (1976) have further suggested that a respiration monitor should be noninvasive, requiring neither surgical intervention nor direct contact with the patient. Obviously, any respiratory monitor must be absolutely safe, producing neither long-term nor short-term damage to the subject.

While the above criteria were developed for apnea monitors used to protect human infants, they are equally applicable to devices used to monitor infant monkeys. If anything, the observation of respiration in infant monkeys is more challenging than it is in human infants. Human neonates cannot roll or creep, while neonatal monkeys can easily move and climb. Thus, a monitor for infant monkeys must be wireless at the very least, since wires can encumber, entangle, and even strangle an unrestrained animal that is capable of coordinated movement.

Several respiration monitors operate either as wireless devices or without contacting the animal at all. Some of these include radio telemetry devices, pressure-sensitive mattresses, and ultrasonic motion detectors. We have tested a wireless impedance pneumograph in our laboratory. It has the advantage of producing both an electrocardiogram and an impedance pneumogram, but it is expensive. Pressure-sensitive mattresses are not only expensive, but they are designed for the motion of large human infants and are susceptible to puncture as well as to other reliability problems (Health Devices, 1974; Franks, Brown, and Johnson, 1976). While ultrasonic motion detectors are usable, those that are readily available operate at 40 kHz, a frequency that is in the auditory passband of the monkey (Pfingst, Heinz, Kimm, and Miller, 1975), and exposure to sound at that frequency may cause undue discomfort to the animal.

We have developed a respiration monitor and apnea alarm that uses a continuous-wave radar system operating at a frequency of 10.525 GHz (Spelman, Kindt, Bowden, Sackett, Spillane, and Blattman, 1975). A similar device has been used in England to monitor human infants (Franks et al., 1976). Our radar system has operated successfully for two years, providing indications of respiratory distress, alarms for prolonged apnea, and an output signal that is a faithful representation of the movements of the chest wall of the infant monkey under observation.

The radar system is safe. The maximum power density at the position at which the animal is observed is 0.74 $\mu W/cm^2$; this is 13,000 times lower than the standard established by the American National Standards Institute, 6,500 times less than the standard established for microwave ovens by the U.S. Bureau of Health, Education and Welfare, and 13 times lower than the most stringent radiation standard in the world, that of the Soviet Union (Guy, 1977).

SYSTEM DESIGN

A radar system consists of a transmitter that radiates a continuous wave of energy at 10.525 GHz, and a receiver that responds to all of the local radio-frequency energy at 10.525 GHz (Figure 1). The radiated energy of the transmitter produces a standing wave whose characteristics are altered by any movement of reflecting structures within the field of the radar system. Changes in the standing wave pattern are sensed by the receiver.

If the transmitter produces a sinusoidally varying electromagnetic plane wave that is perpendicularly incident upon a perfectly conducting plane surface, a standing wave is produced having an electric field strength as given below (Jordan and Balmain, 1968):

$$E(x,t) = 2E_t \sin(\beta x)\sin(\omega_c t) \tag{1}$$

where E_t is the electric field strength of the transmitted wave, t is time, $\beta = 2\pi/\lambda$, λ is the wavelength of the signal, $\omega_c = 2\pi f_c$, and f_c is the frequency of the transmitter.

Equation 1 shows that the electrical field strength of the standing wave is a sinusoid whose amplitude is a function of the distance, x, from the reflector. If the reflector moves, x varies and the amplitude of the standing wave at the receiving antenna varies as well. Let the distance, x, vary sinusoidally:

$$x(t) = x_o + x_1 \sin(\omega_x t) \tag{2}$$

where x_o is the mean distance from the radar to the target, x_1 is the peak change that occurs in distance during a single cycle, $\omega_x = 2\pi f_x$, and f_x is the frequency at which x varies. Let $x_1 \ll \lambda$.

The electric field, E_t will be:

$$E(x,t) = [E_a \cos(k_x \sin\omega_x t) + E_b \sin(k_x \sin\omega_x t)] \sin\omega_c t \tag{3}$$

$E_a = E_t \sin 2\pi x_o/\lambda$, $E_b = E_t \cos 2\pi x_o/\lambda$, $k_x = 2\pi x_1/\lambda$, all constants.

Equation 3 can be composed into a set of sinusoids whose frequencies are sums and differences of the carrier frequency, f_c, and the frequency at which x changes, f_x. If the electric field is received by the receiving antenna, a voltage can be measured at the antenna terminals. That voltage has the form of Equation 3. If the voltage at the antenna is mixed with a voltage whose frequency is that of the carrier, and the output of the mixer is filtered, then it is possible to retrieve a sum of two voltages, one a constant and the other a sinusoid of frequency, f_x:

$$v(t) = V_1(x_1)\sin\omega_x t + V_2 \tag{4}$$

This voltage is amplifed by the intermediate-frequency amplifier. If the intermediate-frequency amplifier is AC coupled, the DC term of Equation 4 is not amplified. The AC term is amplified and used to represent the moving surface that is illuminated by the radar.

The brief analysis given above assumes that the moving surface is plane, perfectly conducting, moving in the direction of propagation, and that the amplitude of movement is much less than a wavelength. In the actual conditions of measurement, the first two assumptions are violated. The moving surface, the body of the monkey, is neither plane nor a perfect conductor. However, studying the simpler analytical case serves to illustrate the principles underlying the operation of the radar. The third assumption is sometimes true and sometimes untrue, while the last assumption is fulfilled. While the motion of the body of the monkey is not always in the direction of propagation of the transmitted wave, the distance of motion is usually less than one-tenth of a wavelength when the animal is at rest. Because the motion of the target is limited, this radar system differs from a true Doppler radar in which the target moves through many wavelengths.

The output signal from the mixer can vary from several hundred microvolts to several millivolts. The gain of the system was designed to vary from 0 to 20,000. Two radars were built in our laboratory. In the first, the bandwidth of the intermediate-frequency amplifier was 0.2-2.0 Hz, while in the second it was 0.2-4.8 Hz. The bandwidth was increased in the second radar to study the detailed frequency characteristics of the respiration signal.

The first radar employed an IMPATT diode source (RACON, Seattle, WA) and a Schottky-barrier mixer (Hewlett-Packard, Palo Alto, CA). The second used a Gunn diode source and Schottky-barrier mixer (Amperex, Hicksville, NY). The power output of the

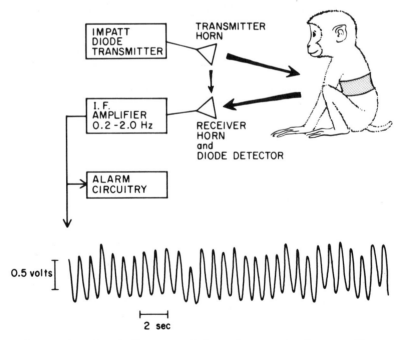

Figure 1.   Block diagram of the radar respiration monitor.

second radar is five times less than that of the first. In addition, the Gunn diode operates with a low-voltage power source, while the IMPATT diode requires a high-voltage source.

Figure 1 illustrates the components of the system: the transmitter, which illuminates the monkey; the receiving antenna, mixer and intermediate frequency amplifier, which detect the changes in standing waves that result from movements of the monkey's body; and the alarm circuits, which alert the laboratory personnel that respiration has ceased.

The apnea alarm is a zero-crossing detector, which produces an output each time the animal breathes, and a timing circuit which produces an output if no zero-crossings are detected within a preset time interval. The interval is adjustable from 0 to 30 sec. The alarm is an audible tone which can be heard through the walls of the monkey nursery. The entire system is small enough (16 x 13 x 10 cm) to be mounted on top of an incubator.

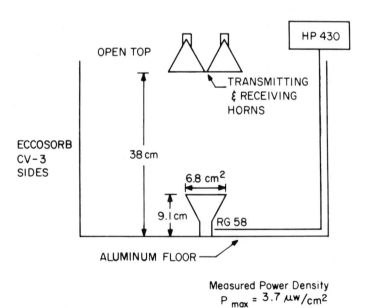

Measured Power Density
$$P_{max} = 3.7\ \mu w/cm^2$$

Figure 2.   Apparatus used to determine the power density produced by the radar transmitter.

TESTS OF THE RESPIRATION MONITOR

The system was tested in several ways: 1) the power output was measured to determine that it would be sufficiently low to allow measurement of respiration in living monkeys; 2) the output of the system was tested when it illuminated a moving target, to assure that there was sufficient output to permit practical use of the system; 3) preliminary measurements were made on infant monkeys; and 4) a study was performed on fetal and neonatal rats to be sure that low-intensity radiation produced no ill effects on the developing organism (Kindt, Bowden, Spelman, and Morgan, 1975).

Figure 2 shows the arrangement of apparatus used to determine the power density at the level of the monkey's body wall. The power density in the figure was recorded from the IMPATT diode transmitting system (Spelman et al., 1975). The corresponding power density for the Gunn diode transmitter is 0.74 $\mu W/cm^2$. The power density that was measured is less than the world's most stringent specification for continuous microwave radiation, the Soviet standard of 10 $\mu W/cm^2$.

Tests were made to determine the response of the radar system to a target that moved in a direction parallel to the direction of propagation of the transmitted wave, with a trapezoidal pattern of 1 Hz. The experimental apparatus is shown in Figure 3. Figure 4 shows the results of tests on a Gunn diode radar respiration monitor. The target moved ±1 mm. The output of the

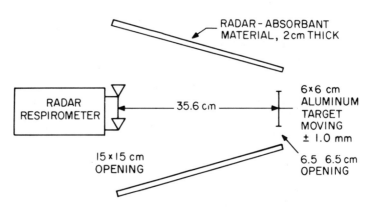

Figure 3.   Apparatus used to measure the response of the radar respiration monitor to a moving metal target.

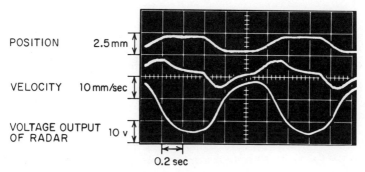

Figure 4.   Output of the respiration monitor to a moving metal
target. The top trace is the position of the target; the
middle trace is the velocity of the target; the bottom
trace is the output of the radar.

radar is a trapezoidal waveform with a peak-to-peak amplitude of
20 v. While the output is distorted, it still gives a reasonable
indication of the movement of the target. The first intermediate
frequency amplifier was used during this test, and the limited
bandpass of that amplifier caused the distortion shown in the
figure.

OBSERVATIONS OF INFANT MONKEYS

Tests were made to determine whether the respiration monitor could obtain reliable results when it was used to monitor infant
monkeys. Monkeys were observed during sleep (Figure 5), during
wakeful rest (Figure 6), and during movement (Figure 7). The
waveforms observed when the monkeys were immobile are nearly
sinusoidal in shape, with insignificant harmonic content beyond the

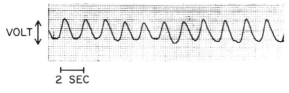

Figure 5.   Output obtained from the monitor when it was used to
observe a sleeping monkey.

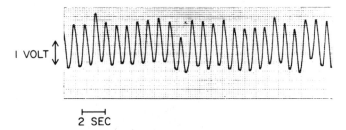

Figure 6.   Output obtained from the monitor when it was used to observe a resting, awake monkey.

third harmonic.   Table I shows the relative power and voltage measured in the first harmonic, as well as in the higher harmonics of the respiration waveform produced by a resting monkey.   The result noted in the ratio of the fourth through the tenth harmonics was obtained by summing the power ratios of the seven harmonics. The result calculated in the ratio of the fourth through tenth harmonics of the output voltage was obtained by computing the square root of the sum of the squares of the voltages, and comparing that value with the magnitude of the fundamental frequency.   The respiration waveform of the resting animal contains insignificant information above the third harmonic.

The output of the radar returns rapidly to normal after the monkey moves grossly.   Figure 7 shows that the return of the signal to a normal respiratory pattern occurs within 2 sec after gross

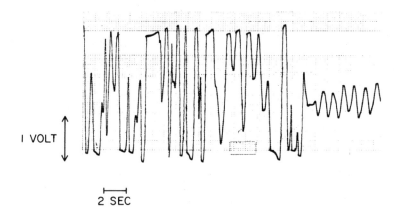

Figure 7.   Output obtained from the monitor when it was used to observe a moving monkey.

TABLE 1.   RATIOS OF THE POWER AND VOLTAGE OUTPUTS BETWEEN THE HARMONICS OF RESPIRATION IN A RESTING MONKEY

Harmonic	Power Ratio	Voltage Ratio
1	1.00	1.00
2	0.05	0.22
3	0.01	0.10
4-10	0.002	0.04

body movements end. During those movements, respiration signals can be observed superimposed on the larger signals caused by body movement. With other circuitry, it may be possible to filter and separate the two signals. We have done some work with a digital filter having a bandwidth of 1.5-2.0 Hz and 10-pole Butterworth characteristics (Javid and Brenner, 1963), and have found that it is possible to filter respiration from movement.

The respiration monitor has been used successfully to detect apnea. Figure 8 shows the result of a measurement made in one animal that had repeated apneic episodes. Not only are the apneas

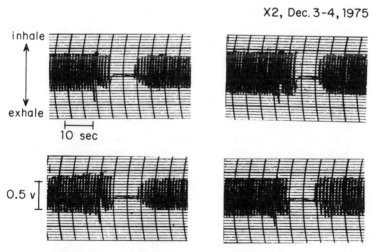

X2, Dec. 3-4, 1975

inhale
exhale
10 sec
0.5 v

Figure 8.   Outputs obtained from the monitor during repeated apneas observed in an infant monkey.

similar in duration, but each is preceded by an additional respiratory anomaly. The change in respiration precedes the 10-sec apnea by 5 or 6 sec in each case. A second animal that exhibited repeated apneas had a similar change in respiratory pattern preceding apnea, but unfortunately the change was not found consistently in this animal. As we gather more data we should be able to determine whether a change in respiratory pattern can be used to alert laboratory personnel to expect and to treat an impending apnea. Other interesting patterns of respiration have been observed with the radar system; they are discussed in Chapter 16.

CONCLUSIONS

We have designed, built, and tested a radar respiration monitor that is safe and reliable for use in infant monkeys. It can be produced easily from components that are commercially available for a cost of less than $300 in unit quantities. The monitor produces an output that is usable for research in the primate nursery, and it provides nursery personnel with a reliable apnea alarm. This monitor is potentially useful as an apnea monitor for human infants.

While the radar monitor has been useful in our nursery, it can be improved. The present system has nulls in the standing wave pattern, but these can be compensated by using dual receivers (Franks et al., 1976). The addition of a narrow-band filter can probably permit the recording of reliable respiration data when the monkey is moving.

The system shows the potential for discriminating central and laryngeal apnea (Guilleminault, Perrita, Souquet, and Dement, 1975). We performed a single experiment on an infant monkey which was lightly anesthetized and intubated. The tracheal tube was alternately opened and closed by means of a three-way stopcock. Changes were observed in the amount of energy contained in the second and third harmonics of the respiratory waveform. While further experiments are necessary to verify these preliminary results, the potential of the monitor is obvious. The discrimination of central and laryngeal apnea would make the monitor useful in the human nursery as well as in the primate nursery.

ACKNOWLEDGMENTS

This research was supported by NIH grants RR00166 and HD08633.

REFERENCES

Franks, C. I., Brown, B. H., and Johnson, D. M.   Contactless respiration monitoring of infants.   Med. biol. Engineer. 14: 306-312, 1976.

Guilleminault, C., Perrita, R., Souquet, M., and Dement, W.   L. Apneas during sleep in infants: Possible relationship with sudden infant death syndrome.   Science 190:677-679, 1975.

Guy, A. W.   Letter to the Human Subjects Review Committee of the University of Washington, May 2, 1977.

Health Devices.   Infant apnea monitors.   Pp. 3-24, Nov.   1974.

Javid, M. and Brenner, E.   Analysis, Transmission, and Filtering of Signals, New York: McGraw-Hill, 1963.

Jordan, E. C. and Balmain, K. G.   Electromagnetic Waves and Radiating Systems, 2nd Ed., Englewood Cliffs, N.J.: Prentice-Hall, 1968.

Kindt, C. W., Bowden, D. M., Spelman, F. A., and Morgan, M.   K. Some developmental and behavioral factors of low-intensity x-band radiation.   USNC/URSI Annual Meeting--Program and Abstracts, B12-5, 1975.

Pfingst, B. E., Heinz, R., Kimm, J., and Miller, J.   Reaction-time procedures for measurement of hearing.   I. Suprathreshold functions.   J. acoust. Soc. Amer. 57:421-430, 1975.

Spelman, F. A., Kindt, C. W., Bowden, D. M., Sackett, G.   P., Spillane, J. W., and Blattman, D. A.   Remote measurement of respiration in infant primates using an x-band Doppler radar. USNC/URSI Annual Meeting--Program and Abstracts, B26-2, 1975.

CHAPTER 16

MONITORING AND APNEA ALARM FOR INFANT PRIMATES: PRACTICAL AND RESEARCH APPLICATIONS

C. W. Kindt and F. A. Spelman

Regional Primate Research Center
University of Washington
Seattle, Washington

INTRODUCTION

The need for a remote respiration monitor became apparent with the increasing number of premature, low-birth-weight, and maternally rejected infant monkeys handled by the Infant Primate Research Laboratory at the University of Washington. Of primary interest was the development of a safe, reliable, contactless device that would free laboratory personnel from the tedious duty of extended observation of one or more acutely ill infants. As a result of this need a radar respiration monitor was developed (Spelman, Kindt, Bowden, Sackett, Spillane, and Blattman, 1975). Initially the monitor was intended to signal prolonged and frequent apneas in the premature infants. However, the radar respiration unit rapidly demonstrated its ability to measure subtle respiration patterns in a variety of disease states as well. We have monitored respiration patterns of infant monkeys that exhibited one or more of the following conditions: prematurity, low birth weight, pneumonia, hyperbilirubinemia, hyaline membrane disease, severe hypothermia, complications arising from post-operative recovery from anesthesia, and trauma. Use of the radar monitor has led to earlier therapeutic intervention and treatment of infants with suspected respiratory compromise than was previously possible through simple observation.

Although the radar unit was not originally designed and built with a particular research application in mind, it has rapidly evolved into an extremely useful research tool. The infant Macaca nemestrina is considered a good model for both upper airway anatomic maturation (Taylor, Sutton, and Lindeman, 1976) and respiratory pathology (French, Morgan, and Guntheroth, 1972) in human infants. Infant M̲. nemestrina also have a higher incidence of Sudden Infant Death Syndrome, as determined by autopsy, than do human infants (Ray and Peterson, 1975). This animal also shows cardio-pulmonary physiologic development that is remarkably similar to that of human infants (Kindt, unpublished data). These facts, coupled with the availability of significant numbers of premature infants and/or newborn animals with compromised respiration, have provided the framework for a number of research projects.

Our research involving the radar monitor has two basic areas of interest. One is further development and sophistication of the device itself so that it can be used with human infants. The other is to further examine the developmental physiologic relations between infant M̲. nemestrina and infant humans.

METHODS

We monitored eight infant M̲. nemestrina determined to be at high risk for respiratory problems, and two full-term normal infants. The high-risk infants ranged in age from 2 hr postpartum to 30 days old, and weighed 273-800 g. The full-term normals weighed 473 and 415 g at birth and were monitored for short periods on days 1, 10, and 20 of life. All animals were housed in a specially modified infant isolette fitted with the radar monitor. Strips of 8-inch-high anechoic material were placed around the inside walls of the isolette to minimize outside interference caused by movement of laboratory personnel. The animals were unrestrained with the exception of two infants that required cardiac monitoring and mechanical ventilation during part of their recovery. The data were collected on a two-channel Brush strip chart recorder and later a four-channel Hewlett Packard FM tape recorder was added to provide a means for more detailed analysis of the respiration patterns. The theoretical, design, and performance characteristics of the radar monitor are detailed in Chapter 15.

The actual methods varied from infant to infant partially because each animal required different amounts of intervention, and partially because our methods evolved simultaneously with development of the radar unit. Animals were determined to be at risk for respiratory problems either by degree of prematurity, obvious trauma (e.g., hypothermia), or by the appearance of apnea. Initially, those animals that required mechanical stimulation of the chest to resolve a prolonged apnea spell were monitored both by the radar unit and by an observer who could stimulate the animal if respiration stopped. An audible alarm was subsequently added and could be set to go off in response to apneic periods of up to 30 sec. This meant laboratory personnel were freed from attending to short, spontaneously resolved apneas.

The data collected on the strip chart were then analyzed by hand for evidence of anomalies in the normal smooth pattern of breathing, as shown in Figure 1. The data collected on magnetic tape were analyzed by a PDP-11 computer for similarities of power spectra.

OBSERVATIONS

We have seen a number of different respiratory patterns while monitoring infant monkeys. Of these, five have particular practical interest and potential research application: apnea, obstructed breathing, Cheyne-Stokes breathing, episodic or periodic breathing, and intermittent sighs.

Apnea

Perhaps the most frequently observed respiratory change in the infant monkeys monitored was apnea. Every monkey, except the normals, exhibited some evidence of apnea. This is not surprising because periodic breathing and apnea occur occasionally in normal human infants (McGinty and Harper, 1974; Steinschneider and Rabuzzi, 1976) and are common in premature or low-birth-

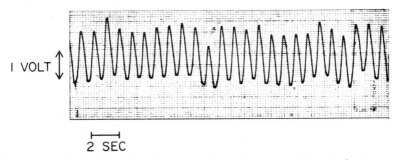

I VOLT

2 SEC

Figure 1.   Normal, smooth respiration from a resting infant monkey. Inspiration is up in all figures.

├──────┤
10 sec   Animal - X-2

Figure 2.   Repeated   patterned   apneas   from   a   low-birth-weight
             monkey.

weight infants (Daily, Klaus, and Meyer, 1969; Shannon, Gotay,
Stein, Rogers, Todres, and Moylan, 1975).   Figure 2 shows a series
of apneic episodes recorded from an immature infant monkey that
weighed less than normal at birth.   The animnal developed apnea
while recovering from the effects of anesthesia used during umbilical hernia repair.   This apneic episode was similar to those seen in
the other animals we monitored.

     Several features of this apneic series are of particular
interest.   One is that one apneic episode seems to trigger subsequent episodes with a fairly predictable period, as it does in human
premature infants (Flanagan, Rowe, Hodson, and Woodrum, 1977).
We also observed this feature, to some degree, in all animals that
exhibited clear apnea.

     Although it is not shown in this figure, ECG and respiration
were monitored simultaneously in several infants.   These simultan-
eous recordings reveal that another feature of these apneic episodes is a decrease in heart rate approximately 2-3 sec after the
onset of apnea.   In no instance did heart rate fall prior to an apneic
spell.   This pattern is also characteristic of human infants
(Chernick, Heldrich, and Avery, 1964; Rowe, Flanagan, Hodson, and
Woodrum, 1977).   Rowe and Flanagan found a heart rate decrease
to occur 9.7 ±4.0 sec after onset of apnea in human infants.

     A third feature of this series of apneic spells is that
approximately 5-6 sec prior to each apneic episode there is a
respiratory change much like a shortened breath (arrows on figure).
A similar pattern was seen in one other animal.   To our knowledge,
this phenomenon has not been reported in the human literature.

Obstruction

     An obstructed or "loaded" breathing pattern occurs almost as
frequently as apnea.   Figure 3 shows a typical bout of obstructed

2 sec   Animal – B W

Figure 3.   Pattern of upper airway obstruction from a severely hypothermic infant monkey.

breathing. This pattern seems to be the result of partial plugging of upper airways and nasal passages or of a marked increase in upper airway resistance. Close observation of the infant during one of these spells of obstructed breathing reveals a clear asynchrony between chest wall and abdominal movements. In a recent report to a national conference on Sudden Infant Death Syndrome (SIDS), Bryan and Bryan (1977) observed that a similar chest wall and abdominal disorganization could be produced by adding an elastic load to a human infant's airway. This disorganization can lead to a signficantly lower tidal volume and minute volume, which in turn could produce episodes of hypoxia and hypercarbia. Although the apparent size of the respiratory excursions in the figure seem large, it is the motion of the chest wall that is being measured and not tidal volume.   When breathing is partially obstructed, the respiratory movements tend to be rapid and disorganized rather than smooth and slow as seen in Figure 1, thus producing relatively higher output voltage from the radar.

This particular infant was shipped in a poorly heated container and it arrived at the Infant Laboratory with a rectal temperture of 81° F. The obstructed breathing pattern started shortly after arrival and we began monitoring the animal immediately.   It subsequently developed apnea and was placed on 100% oxygen coupled with postural drainage and suction. Respiration monitoring and therapy were continued for two days during which time the infant regained a normal smooth respiration pattern and recovered fully.

Cheyne-Stokes

A third type of pattern resembles Cheyne-Stokes breathing, which was described as early as 1818 by J. Cheyne and is characterized by a changing respiratory rate and a decreasing tidal volume as shown in Figure 4. Often this pattern is followed by a period of apnea and a reverse pattern of rate and depth of breathing. Cheyne-Stokes breathing can be caused by at least three different mechanisms: 1) increased circulation time secondary to heart failure, 2) instability or perhaps immaturity of the central or peripheral neural respiratory control system, and 3) intermittent upper airway obstruction (Cherniack and Longobardo, 1973). Although this breathing pattern is generally associated with older and/or obese adult humans, we have seen it repeatedly in the infants monitored. Chernick et al. (1964) suggest that the premature baby with periodic breathing resembles the adult with Cheyne-Stokes respiration. Since increased circulation time due to heart failure is an unlikely etiology for this pattern of respiration in infants, it seems probable that prematurity, neural immaturity or insult, or intermittent upper airway occlusion may be responsible for this pattern. This animal was of low birth weight, probably premature, and later died of hyaline membrane disease. The infant seemed to improve after several days of increased inspired O_2 therapy and although it never did breathe "normally," it was taken off oxygen, monitoring was discontinued, and it was put into a regular cage where it ultimately died some days later. The Cheyne-Stokes breathing pattern was also observed in two other infants. One was of very low birth weight and the other had an elevated serum bilirubin of more than 13 mg%.

Episodic or Periodic Breathing

An episodic breathing pattern has been somewhat difficult to characterize. A typical example is shown in Figure 5. This animal was taken by Caesarean section at 142 days gestation (26 days premature) and was one of the first animals to be monitored. It

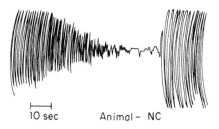

10 sec   Animal - NC

Figure 4.   Cheyne-Stokes breathing pattern in an animal with hyaline membrane disease.

had apneic episodes similar to those in Figure 2. The infant was sacrificed as part of another experimental protocol and no other respiratory complications were apparent. Even though this breathing pattern does not precisely fit the periodic breathing defined by either Chernick et al. (1964) or Rigatto and Brady (1972), it has been included because of its repeated occurrence in several other animals that were monitored. Although this pattern is not associated with increased movement or with a particular sleep state as far as we can determine, it is entirely possible that further observation will show it to be related to active or rapid eye movement (REM) sleep. Analysis of respiratory patterns in human infants during sleep indicates that during REM sleep, ventilation becomes significantly more irregular and erratic than during non-REM sleep (Prechtl, Akiyama, Zinkin, and Grant, 1968; Hathorn, 1974). Chernick et al. (1964) have suggested that rapid changes in the newborn breathing pattern may be a feature of immaturity of the developing nervous system. If this is true, these rapid changes in pattern may be more pronounced in infants born or taken prematurely and are shown by the episodic patterns in Figure 5.

Intermittent Sighs

A respiration pattern that is infrequently seen, but is nonetheless interesting, seems to be a sigh in that the long downward swings of respiration are the result of rapid expiration (Figure 6). The animal that produced this pattern showed symptoms of hyaline membrane disease as well as breathing patterns similar to those of an earlier animal that died of hyaline membrane disease. The animal was premature, had a low birth weight, and became hypothermic 1 hr after birth. It was put on a small-animal

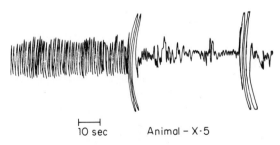

10 sec   Animal – X-5

Figure 5.   Episodic, erratic or periodic pattern of respiration in an infant delivered by Caesarean section at 142 days gestation (26 days premature).

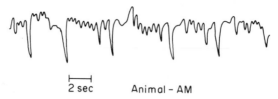

2 sec   Animal - AM

Figure 6.   Intermittent sighs from a low-birth-weight, premature
infant monkey.

ventilator with 100% O_2 and had an umbilical intravenous catheter
for direct feeding and fluids.   We removed the ventilator after
about 10 hr of therapy and began to monitor respiration with the
radar unit.   After several weeks of increased inspired oxygen
treatment and continuous respiration monitoring, the animal recovered fully.   The etiology and mechanism for this pattern are
purely speculative at this point.   The speculation centers around
two possibilities. One is that there is a buildup of CO_2 in the blood
due to the small chest excursions before the sigh.   The sigh then
would help eliminate part of this CO_2.   The other possibility is that
some air trapping occurs in the lung and only when the amount of
air trapped reaches some critical volume or pressure is it released
in the form of an expiratory sigh.

CONCLUSIONS

Currently the radar respiration monitor has numerous practical applications. It is small, safe, reliable, and potentially afford-
able in quantity for nearly any primate nursery situation.   The
output from the radar can be used in a variety of ways to help
decrease infant mortality in a nursery and to free laboratory
personnel from hand-collection of certain types of diurnal data or
from long periods of observation of acutely ill animals. Depending
on how the output from the radar is displayed, it can be used to
simply alert laboratory personnel of a prolonged apnea spell in an
infant needing mechanical intervention, or it can be used to detect
early, subtle changes in respiration patterns that may indicate
airway obstruction requiring simple examination and suction of the
throat.   Changes in rate of respiration can quickly be seen if the
output is fed to a device that gives a constant digital display.
These rate changes are often associated with fluid buildup in the
lung as the result of aspiration pneumonia, prolonged hypoxia,
nervous system insult or trauma, or infection (Dr. L. R. O'Grady,

personal communication). Gross changes in the pattern of respiration are also easily recognized and even though there currently is no clear correlation between a specific breathing pattern and a specific pathology, certain gross changes often indicate disease onset. Early recognition and treatment are obvious benefits of this type of monitoring. Routine screening of a particular infant population becomes possible as well as periodic collection of respiration or activity information.

The potential research applications of this monitor are innumerable. Currently, work is in progress to do spectral analysis on the repeated pre-apnea patterns seen in Figure 2. By such analysis, early warning of pending respiratory arrest may be possible. This analysis may also make it possible to identify those infants that are at particularly high risk for respiratory problems. Spectral analysis may also reveal correlation of a particular respiratory pattern with a specific pathologic condition. It may prove that a pneumonia pattern is unlike that of atelectasis which is unlike central nervous system immaturity or damage. If this type of analysis proves successful, a glossary of respiratory patterns and frequency spectra will be developed for both normal and abnormal infants, which could then be used as the basis for extensive modeling of human infant respiratory pathology.

Recent experiments with the radar monitor have revealed differences between central and peripheral apnea and normal respiration in infant monkeys (Spelman, Sutton, and Kindt, unpublished data). Central apnea, i.e., total cessation of respiratory movement, is due to a lack of nervous output from central respiratory centers. Peripheral apnea, the obstruction of air flow into and out of the lungs with continuing respiratory movements, could be caused, for example, by closure of the glottis due to laryngospasm. Differentiation between these types of apnea is important because they are both thought to occur in SIDS (Steinschneider, 1970). Any monitor that might be used to detect breathing cessation of a child at risk for SIDS should be able to discriminate between normal respiratory movements and respiratory movements against a closed glottis.

A major assumption we have made in this chapter is that the changes observed in respiratory patterns are evidence of underlying pathology. However, the records of Dening and Washburn (1935), who were among the first to record respiratory patterns from normal, apparently healthy infants, show several patterns similar to those we observed in premature or sick newborn monkeys. Periodic breathing, short apneas, a modified type of Cheyne-Stokes breathing, and occasionally a very irregular pattern were among their recordings. On the other hand, Johnson et al. (1964) showed that many of these patterns are very similar to those seen in newborn lambs with experimentally produced respiratory distress syndrome. They showed apnea, bradypnea, labored respiration, rib retraction, and other respiratory changes in their model. Since our monitoring experience with healthy, full-term infant monkeys is limited, it is impossible to be certain that the patterns we have presented are necessarily of pathologic origin. We have not seen these patterns thus far in any full-term, healthy infant monkey. The bulk of the human literature on this subject clearly indicates that the types of respiratory patterns we have presented are closely associated with a variety of newborn pathology, including SIDS, idiopathic respiratory distress syndrome, pneumonia, hyaline membrane disease, prematurity, and lung obstruction.

ACKNOWLEDGMENTS

This work was supported by NIH grants HD02274, RR00166, and HD08633.

REFERENCES

Bryan, M. H. and Bryan, A. C. Response of infants to respiratory loads. The First Annual Workshop of Sudden Infant Death Syndrome Grantees and Contractors. NICHHD Perinatal Biology and Infant Mortality Branch. Publication #RO 1 HD 07826-02, 1977.

Cherniack, N. S. and Longobardo, G. S. Cheyne-Stokes breathing: An instability in physiologic control. New Engl. J. Med. 288:952-957, 1973.

Chernick, V., Heldrich, F., and Avery, M. E. Periodic breathing of premature infants. J. Pediat. 64:330-340, 1964.

Daily, W. J. R., Klaus, M., and Meyer, H. B. P. Apnea in premature infants: Monitoring, incidence, heart rate changes, and an effect of environmental temperature. Pediatrics 43:510-517, 1969.

Dening, J. and Washburn, A. H. Respiration in infancy. Amer. J. Dis. Child. 49:108-124, 1935.

Flanagan, W. J., Rowe, J. C., Hodson, W. A., and Woodrum, D. E. Apnea immediately following apnea in premature infants. Ped. Res. Abstract #969, April, 1977.

French, J., Morgan, B., and Guntheroth, W. Infant monkeys: A model for crib death. Amer. J. Dis. Child. 123:480-484, 1972.

Hathorn, M. K. S. The rate and depth of breathing in new-born infants in different sleep states. J. Physiol. 243:101-113, 1974.

Johnson, J. W., Salem, E., and Holzman, G. Experimental induction of respiratory distress in fetal and newborn lambs. Amer. J. Obstet. Gynec. 80:481-497, 1964.

McGinty, D. and Harper, R. M. Sleep physiology and SIDS: Animal and human studies. In: SIDS, 1974. Proceedings of the Francis E. Camps International Symposium on Sudden and Unexpected Deaths in Infancy, R. R. Robinson (Ed.), Canadian Foundation for the Study of Infant Deaths, 1974.

Prechtl, H. F. R., Akiyama, Y., Zinkin, P., and Grant, D. K. Polygraphic studies of the full-term newborn. I. Technical aspects and qualitative analysis. Pp. 1-21 in: Studies in Infancy, R. Mac Keith and M. Bax (Eds.), London: Heinemann Medical, 1968.

Ray, C. G. and Peterson, D. R. Surveillance of newborn Macaca nemestrina in a breeding colony for SIDS and viral infection: Joint report for U. of Washington Regional Primate Research Center and Children's Orthopedic Hospital and Medical Center, 1975. (Unpublished).

Rigatto, H. and Brady, J. P. Periodic breathing and apnea in preterm infants. I. Evidence for hypoventilation possibly due to central respiratory depression. Pediatrics 50:202-218, 1972.

Rowe, J. C., Flanagan, W. J., Hodson, W. A., and Woodrum, D. E. The relationship between apnea and heart rate in premature infants. Ped. Res., Abstract #1019, April, 1977.

Shannon, D. C., Gotay, F., Stein, I. M., Rogers, M. C., Todres, I.D., and Moylan, F. M. B. Prevention of apnea and bradycardia in low-birthweight infants. Pediatrics 55:589-594, 1975.

Spelman, F. A., Kindt, C. W., Bowden, D. M., Sackett, G. P., Spillane, J. W., and Blattman, D. A. Remote measurement of respiration in infant primates using an x-band Doppler radar. USNC/URST Annual Meeting, Program and Abstracts, B26-32, 1975.

Steinschneider, A. and Rabuzzi, D. D. Apnea and airway obstruction during feeding and sleep. Laryngoscope 86:1359-1366, 1976.

Steinschneider, A. Possible cardiopulmonary mechanisms in sudden infant death syndrome. In: Sudden Infant Death Syndrome, A. B. Bergman, J. B. Beckwith, and C. G. Ray (Eds.), Seattle: University of Washington Press, 1970.

Taylor, E. M., Sutton, D., and Lindeman, R. C. Dimensions of the infant monkey upper airway. Growth 40:69-74, 1976.

CHAPTER 17

SURVEY OF NEONATAL AND INFANT
DISEASE IN Macaca nemestrina

W. R. Morton, W. E. Giddens, Jr., and
J. T. Boyce

Regional Primate Research Center
University of Washington
Seattle, Washington

Expanding use of nonhuman primates as animal models for human disease requires a more thorough understanding of the natural disease spectrum of each species used in research. Anver et al. (1973) described the major pathological findings in neonates of nine species at the New England Regional Primate Research Center. The present report describes the major pathological findings in 63 pigtail macaque (Macaca nemestrina) perinates, neonates and infants at the Infant Primate Research Laboratory of the Child Development and Mental Retardation Center and the Regional Primate Research Center at the University of Washington between 1973 and 1977. The animals were between 1 day and 1 year old at death.

Complete necropsy was conducted on 52 of the 63 animals. Causes of death in the remaining 11 animals were determined by clinical evaluation and gross necropsy only.

At the time of gross necropsy, samples of tissue from all parenchymal organs were collected and fixed in 10% phosphate-buffered formalin. After paraffin-embedding, 5-μm sections were cut and stained with hematoxylin and eosin for routine light microscopic examination. Where indicated, periodic acid Schiff, gram, giemsa, and other special stains were done.

Table 1 lists major pathological findings considered to be the primary cause of death. In several cases the primary cause of death was not conclusively determined, so more than one pathological finding is listed. Sex, age at death, and birth weight are also shown.

227

TABLE 1. CAUSES OF DEATH IN 63 INFANT M. nemestrina

Monkey	Sex	Age*	Birthweight (grams)	Cause of Death
Perinates (0-7 days)				
T76346	M	1	167	premature delivery; pulmonary atelectasis
M76287	M	2	595	dystocia; respiratory distress syndrome
A73-82	F	2	503	bacterial cellulitis--severe; pneumonia-serofibrinous, interstitial
A73-36	M	3	376	inhalation pneumonia--acute
M-73137	F	3	547	brain hemorrhage; pulmonary atelectasis
A74-37	F	3	325	cerebellar hematoma; diffuse meningeal congestion; bronchopneumonia--purulent, severe; inhalation of amniotic fluid; interventricular septal defect
T76175	F	3	312	premature delivery; pulmonary atelectasis
T75006	M	5	292	inhalation pneumonia
A75-11	M	5	460	enterocolitis
A77-56	F	5	--	skeletal muscle necrosis--severe
M73180	F	7	390	Shigellosis (enteritis)
A74-7	F	7	320	inhalation bronchopneumonia--purulent, bacterial, severe
A75-92	M	7	308	peritonitis--fibrinopurulent, severe

*Age in days for perinates and neonates; in months for infants

Monkey	Sex	Age	Birthweight (grams)	Cause of Death
Neonates (8-28 days)				
A-75-58	F	9	390	necrotizing enterocolitis
T76177	M	10	253	idiopathic respiratory distress (hyaline membrane disease)
M73199	M	10	558	pneumonia; cerebral edema
A76-64	F	14	421	interstitial pneumonia--acute
A77-72	M	16	385	skeletal muscle coagulation necrosis; acute necrosis of large intestine with severe neutrophilic infiltration
A76-81	F	22	452	adenoviral pneumonia; lobar emphysema secondary to congenital malformation
A76-12	M	24	572	inhalation pneumonia--acute, fibrinopurulent
T75112	M	25	480	acute airway obstruction--ingesta
A76-25	M	25	450	interstitial pneumonia--acute
Infants (1-6 months)				
A75-62	M	1	440	myocardial necrosis--focal
A76-5	F	1	273	peritonitis--fibrinopurulent, acute
T77068	M	1	417	inhalation pneumonia
A74-16	M	1.25	466	inhalation pneumonia--fibrinopurulent, severe

Monkey	Sex	Age	Birthweight (grams)	Cause of Death
A74-78	M	1.25	407	torsion ileal necrosis, hemorrhage--acute
A75-43	M	1.5	5-2	interstitial pneumonia; atelectasis
A75-52	F	1.5	246	necrotizing enterocolitis
A75-51	M	1.5	549	intersititial pneumonia; bacteremia
A73-2	M	2	533	pneumonia--fibrino-purulent, acute
A73-6	F	2	435	inhalation pneumonia--acute; bronchopneumonia--fibrinopurulent
A73-13	M	2	450	inhalation pneumonia-fibrinopurulent
A73-19	M	2	520	inhalation pneumonia--acute; bronchopneumonia--fibrinopurulent
A75-8	M	2	319	inahlation pneumonia--severe
A75-48	M	3	453	adenoviral pneumonia; inhalation pneumonia-bacterial, firinopurulent, pleuritis
A75-55	M	5	490	pneumonitis, interstitial atelectasis
A77104	F	5	403	pulmonary edema--acute, diffuse
A75-7	F	5.5	495	inhalation pneumonia; tubular nephritis--purulent
A75-23	F	6	330	necrotizing, ulcerative enteritis; peritonitis--fibrinopurulent, acute
A46-47	M	6	375	pneumonia--interstitial, subacute

Monkey	Sex	Age	Birthweight (grams)	Cause of Death
Infants (6-12 months)				
A75-66	F	7	448	proliferative stomatitis
A77	M	7	494	proliferative stomatitis
A74-18	F	8	545	inhalation pneumonia--acute; bronchopneumonia--purulent
A75-21	F	8	469	glomerulonephritis
A76-14	F	8	495	proliferative stomatitis; bronchopneumonia; multifocal hemorrhage
T74233	F	9	413	gastric dilatation--acute
A75-69	F	9	424	proliferative stomatitis
A76-13	F	9	447	proliferative stomatitis; colitis--chronic, severe
A76-20	F	9	425	proliferative stomatitis
A77-32	F	9	469	proliferative stomatitis
A76-18	F	10	500	colitis--purulent, severe
A75-41	M	10.5	477	cause undetermined
A75-78	M	11	360	colitis--catarrhal
A76-21	M	11	516	pulmonary atelectasis
M74163	M	12	526	interstitial pneumonia--acute
A75-57	F	12	324	enterocolitis--subacute; pneumonia--interstitial, diffuse
A75-85	M	12	491	glomerulonephritis; tubular necrosis
A76-51	M	12	431	interstitial nephritis
A76-42	F	12	506	meningoencephalitis--suppurative, hemorrhagic, acute

Monkey	Sex	Age	Birthweight (grams)	Cause of Death
A76-56	M	12	572	segmental enterocolitis
A76-87	M	12	580	granulomatous, nodular peritonitis
A77-22	M	12	340	proliferative stomatitis

TABLE 2.   PRIMARY PATHOLOGICAL FINDINGS BY SYSTEM
IN M. nemestrina

System	No. of primary pathological findings
Perinates (0-7 days)	
Respiratory	9
Gastrointestinal	2
Central nervous	2
Cardiovascular	1
Musculoskeletal	1
Other	1
Neonates (8-28 days)	
Respiratory	7
Gastrointestinal	2
Musculoskeletal	1
Infants (1-6 months)	
Respiratory	14
Gastrointestinal	3
Cardiovascular	1
Other	1
Infants (6-12 months)	
Respiratory	5
Gastrointestinal	13
Genitourinary	3
Central nervous	1
Other	1
No significant pathological findings	1

Table 2 presents the primary pathological findings by system and age category. Table 3 shows total numbers of pathological findings by system and age category.

Repiratory disease was the most common finding in infants under 6 months of age. (It was also most commonly found in that age group: 48 of 60 respiratory lesions were in the 0-6 months group.) Inhalation pneumonia was the most common type of respiratory involvement; it predominated in the perinatal and neonatal groups, but was seen throughout the entire age group. Improved management techniques for premature, low-birth-weight, and high-risk infants have resulted in a significant decrease in the occurrence of inhalation pneumonia in our nursery.

Intrauterine pneumonia, reported as a common finding by Anver et al. (1973), was not a significant process in this survey, although it is an important cause of neonatal mortality in M. nemestrina (R. A. Price, personal communication). It did not occur in our sample because most of our newborns come from the Primate Field Station at Medical Lake, 300 miles distant, and perinates that die within the first 24 hr do not reach our nursery in many cases. Price et al. (1973) found that all infants subjected to intrauterine distress in their study died in the first 24 hr of life. If we assume that this same process occurs in M. nemestrina, then deaths due to intrauterine pneumonia would not be reflected in our study. This critical time period should be further evaluated to determine the incidence of this condition in M. nemestrina.

TABLE 3.   TOTAL PATHOLOGICAL LESIONS BY SYSTEM AND AGE CATEGORY IN M. nemestrina

System	Total pathological lesions				
	0-7days	8-28days	1-6mo	6-12mo	Total
Respiratory	12	10	26	12	60
Gastrointestinal	4	5	12	31	52
Genitourinary	0	2	3	14	19
Musculoskeletal	1	1	2	0	4
Haematopoietic and lymphatic	4	2	8	19	33
Central nervous	3	1	5	4	13

The nonhuman primate nursery is becoming a valuable resource for spontaneously occurring neonatal disease conditions that may correspond to human infant disease. Several conditions have been recognized in our laboratory. Two cases of adenoviral pneumonia included in this sample are described in detail elsewhere (Boyce, Giddens, and Valerio, 1978). One of those cases also had a primary diagnosis of congenital lobar emphysema. Cyst-like development at 22 days was sufficient to displace the entire thoracic content. Radiographs at days 1 and 22 postpartum confirm the presence of this congenital cyst. It cannot be determined whether this was a spontaneously occurring lobar emphysema, perhaps the result of fetal intrauterine adenoviral infection. Adenoviral pneumonia has been experimentally induced by fetal inoculation in rhesus macaques (M. mulatta) (Moe, Coburn, Espera, and Schwartz, 1978), and it might similarly be possible to study this disease in M. nemestrina.

Idiopathic respiratory distress syndrome (hyaline membrane disease) has been documented in M. nemestrina (Murphy, Palmer, Prewitt, Young, Standaert, Morgan, and Hodson, 1976). Although only one case was reported in this study, the incidence has been shown to be much higher in premature M. nemestrina, especially those with gestation lengths less than 145 days. We are currently studying M. nemestrina as an animal model for this important human infant respiratory condition.

Neonatal necrotizing enterocolitis was diagnosed in three cases, one of which was judged to be a premature infant based on birth weight. In human cases premature infants are more commonly affected, particularly those at extremely low birth weights (Hopkins, Sould, Stevenson, and Oliver, 1970). The cardinal clinical signs reported for human cases were monitored in our monkeys, including vomiting, temperature instability, abdominal distention, and gastrointestinal bleeding. Bell et al. (1971) reported that 90% of their human infant cases showed at least one clinical sign during the first 10 days of life and more frequently between days 2 and 5. Our three monkeys were 5, 9, and 45 days. Pneumotosis intestinalis was a common feature in all three cases. This condition can be detected by radiography and should be checked whenever cases of abdominal distress are presented.

In infants over 6 months of age we found a marked drop in the incidence of respiratory disease. The most common processes in this age group involved the gastrointestinal system. We observed eight cases of proliferative stomatitis, a disease process that seems to be of viral origin with clinical signs including oral mucosal erosions, particularly of the dorsal aspect of the tongue, catarrhal colitis, and lymphoid hyperplasia. A more detailed description of the clinical and pathological features is in preparation.

Although the total number of infants in this survey is small, the information is meaningful because it is restricted to one primate species. The distribution of pathological lesions is consistent with earlier reports (Anver, Hunt, and Price, 1973; Price, Anver, and Hunt, 1973) when one considers the structure of the infant population in this nursery. Several human disease models have been shown to occur in M. nemestrina, and the value of the nonhuman primate nursery as a major resource for neonatal disease studies should become increasingly apparent as use of these facilities is intensified.

ACKNOWLEDGMENT

This survey was supported by NIH Grant RR00166 from the Division of Research Resources.

REFERENCES

Anver, M. R., Hunt, D. R., and Price, R. A. Simian neonatology. II. Neonatal pathology. Vet. Path. 10:16-36, 1973.

Bell, R. S., Seahorn, C.B., and Stevenson, J. K. Roentgenologic and clinical manifestations of neonatal necrotizing enterocolitis; experience with 43 cases. Amer. J. Roentgenol. 112:123-134, 1971.

Boyce, J., Giddens, E., and Valerio, M. Simian adenoviral pneumonia. Amer. J. Path. 91:259-276, 1978.

Hopkins, G. B., Sould, W. E., Stevenson, J. K., and Oliver, T.K., Jr. Necrotizing enterocolitis in premature infants; a clinical and pathologic evaluation of autopsy material. Amer. J. Dis. Child. 120:229-232, 1970.

Moe, J.B., Coburn, B. I., Espera, C., and Schwartz, L. W. Pathogenesis of adenoviral pneumonia in foetal rhesus monkeys. Lab. Invest. 38:28, 1978.

Murphy, J., Palmer, S., Prewitt, J., Young, J., Standaert, T., Morgan, T., and Hodson, W. A. Hyaline membrane disease (HMD) in the premature monkey. Amer. Rev. resp. Dis. 113:44a, 1976.

Price, R. A., Anver, M.R., and Hunt, R. D. Simian neonatology. III. The causes of neonatal mortality. Vet. Path. 10:37-44, 1973.

IV. HOUSING AND SOCIAL DEVELOPMENT

CHAPTER 18

FACTORS INFLUENCING SURVIVAL AND DEVELOPMENT
OF Macaca nemestrina AND Macaca fascicularis INFANTS
IN A HAREM BREEDING SITUATION

J. Erwin

Department of Psychology
Humboldt State University
Arcata, California

Several large breeding colonies are committed to harem or large group breeding situations, and the success of these programs clearly depends on the survival and development of the offspring. Most of the infants born at the Primate Field Station of the Regional Primate Research Center at the University of Washington are born into harem groups, but a few enter all-female or multi-male groups. Harem groups typically contain a single adult male, 8-13 adult females, and the unseparated offspring of females in the group. In many cases infants are not separated from their mothers at birth to be reared in nurseries, but are instead reared for some time in their natal or other groups. The time at which infants are separated from their mothers ranges from less than 3 months to a little more than 1 year of age, but most infants are removed from their natal groups at 5-7 months. This report focuses on survivorship of two macaque species, the pigtail macaque (Macaca nemestrina) and the crabeating macaque (M. fascicularis), under essentially identical circumstances in harem groups, and on some risk factors that operate contrary to survival in this situation.

SURVIVORSHIP OF INFANTS

Methods

Data on infant mortality (1967-1974 for M. nemestrina and 1967-1976 for M. fascicularis) were retrieved from computerized colony records and individual animal charts, including necropsy and pathology reports. Only live-born infants were included. Animals that left the colony or were involved in intrusive experimental procedures were treated as survivors up to the time they left the colony or received research treatments; at the appropriate time such individuals were removed from the data base and it was assumed that, had they remained in the colony, their chances of survival would have been the same as for those that remained. To the extent that assignment of subjects to experiments may have been non-random, the latter assumption may have been violated. There were potentially biasing influences such as retention of

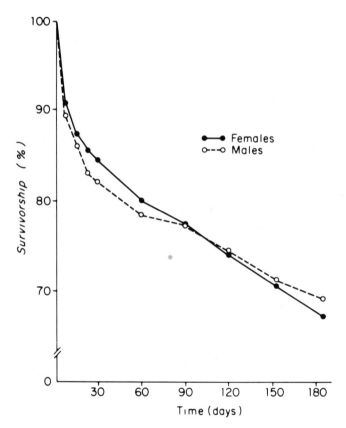

Figure 1.   Survivorship curves for male and female live-born M̲. nem̲estrina̲, 1967 - 1974.

animals that were heavier than average at birth for breeding stock, assignment of light-weight animals to perinatal biology projects, and removal of light-weight or prematurely born infants to the nursery for intensive care.

Infant mortality was defined as death occurring from immediately following live birth to 183 days of age (Hird, Henrickson, and Hendrickx, 1975). Infants found dead the morning after birth (virtually all infants were born at night) were usually given a lung float test to determine whether they had breathed. Neonatal deaths were those occurring within the first 30 days; post-neonatal deaths were those occurring between 31 and 183 days.

Results

During the periods surveyed, 1,174 M. nemestrina and 175 M. fascicularis live births were recorded at the Primate Field Station.

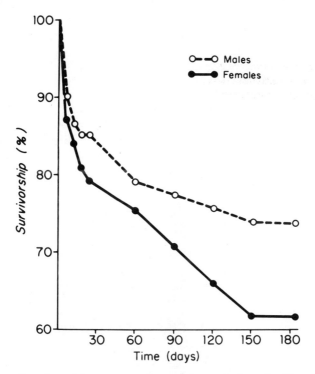

Figure 2.   Survivorship curves for male and female live-born M. fascicularis, 1967 - 1976.

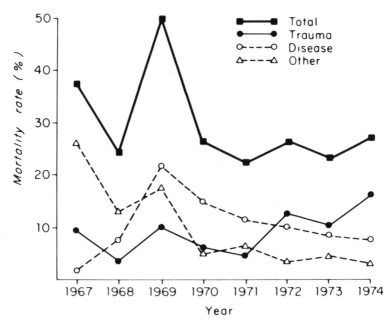

Figure 3a. Infant M. nemestrina mortality rates and cause of death according to year of birth in the colony.

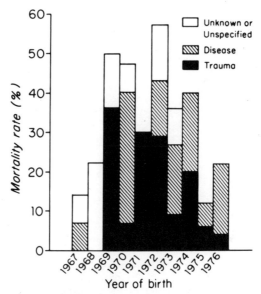

Figure 3b. Infant M. fascicularis mortality rates and cause of death according to year of birth in the colony.

Of the M. nemestrina infants, 17 were of undetermined or unreported sex; of the remaining 1,157 infants, 598 (51.69%) were males and 559 (48.31%) were females. Of the M. fascicularis infants, 93 (53.14%) were males and 82 (46.86%) were females. Thus, the ratio of males to females was 1.07 for M. nemestrina and 1.13 for M. fascicularis.

The infant mortality rate for M. nemestrina was 316/1,072 (29.48%) and for M. fascicularis, 43/141 (30.5%). The neonatal mortality rates were 177/1,126 (15.72%) for M. nemestrina and 28/150 (18.67%) for M. fascicularis, while the respective post-neonatal period rates were 139/895 (15.53%) and 15/113 (13.27%). Survivorship curves for these two species are shown in Figures 1 and 2. There was a noticeable sex difference in survivorship of M. fascicularis, but not M. nemestrina. For comparative follow-up life tables for pigtail macaques, see Dazey and Erwin (1976), and for crabeating macaques, see Erwin (1977a).

Figures 3a and 3b compare mortality rates by year of birth for these two species. For M. nemestrina, infant mortality exceeded 40% only in 1969, while for M. fascicularis it exceeded 40% in 1969, 1970, 1972, and 1974. The causes of death fit into two major classes: disease and trauma. The most common diseases were pneumonia and gastroenteritis; nearly all cases of trauma involved bites (Figure 4).

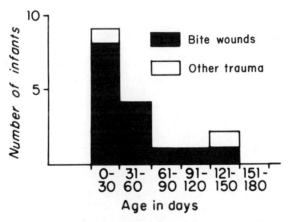

Figure 4.   Infant mortality due to trauma according to age of infant and type of trauma.

Infant deaths due to trauma were particularly common in newly formed groups and in groups with a new resident male. Several cases of males killing or injuring infants were observed. In one series of events, a M. nemestrina infant was injured shortly after an unfamiliar adult male was introduced into a group. The male grabbed the infant and bit its head. The infant and its mother were immediately removed for treatment. About 6 weeks later the mother bit the infant's face while she was being caught from an individual cage. A similar pattern has often been observed when females with infants were being captured from harem groups or individual cages. It is not uncommon for a M. nemestrina female to open her mouth as if to bite her infant's head when she is being pursued. After the infant recovered from the second trauma it was placed along with its mother in a different harem group and was immediately killed by the resident male.

Group instability is clearly a major risk factor with regard to infant trauma. In 1969, for example, 14 M. fascicularis were born into the colony. Of these, 7 were born into stable groups and all 7 survived the first 6 months of life. The other 7 infants were born into a newly formed group and all 7 infants died within 100 days of birth, 5 of bite wounds. (The causes of death for the other 2 infants were not recorded.) In most cases in which males killed infants, the victims were not sired by the perpetrator, but there were some exceptions. More infants were killed during the neonatal period than during any subsequent period. For example, 8 infants were killed during the first month, 4 during the second month, 1 each in months 3, 4, and 5, and none during the sixth month.

Discussion

In both species surveyed here, somewhat more male than female infants were born. Valerio (1969) also reported an excess of male births for M. mulatta, but Hird et al. (1975) and Drickamer (1974) reported a slight preponderance of female offspring for that species. There is some evidence (Dazey and Erwin, 1976) that sex differences in prenatal wastage may account for the sex ratio of live-born offspring. Among both M. nemestrina and M. fascicularis, more male than female fetuses are spontaneously aborted, but more female than male infants are stillborn. There is compelling evidence that prenatal stress can increase the incidence of spontaneous abortion (Chapter 2), and it is conceivable that sex-differential vulnerability could interact with housing conditions and colony procedures to influence sex-ratio outcomes.

The ages at which infants died were similar across species. Of those that died, 41.86% of M. fascicularis and 39.73% of M. nemestrina died within the first week, and 53.49% of M. fascicularis and 51.66% of M. nemestrina died by the end of the second week. Corresponding figures for M. mulatta at the California Primate Research Center for the 2-week period were 34.96% and 52.03% (Hird et al., 1975). Thus, for all three species, more than half the infants that died at less than 6 months of age died within the first 2 weeks. Difficult labor with consequent cerebral hemorrhage, premature birth with subsequent respiratory distress, and trauma due to bite wounds in unstable social groups accounted for most of the deaths in all three species. Intensive nursery care might have saved many of these infants had they been identified as being at risk (cf. Chapter 13). In the case of trauma deaths, most could have been avoided by maximizing group stability and by removing pregnant females from unstable groups before delivery.

Traumatic death of infants has been noted in free-ranging macaque groups including M. mulatta (Carpenter, 1942), M. sylvana (Burton, 1972), and other Old World monkeys such as the red-tailed monkey, Cercopithecus ascanius (Struhsaker, in press). The most notorious cases of infanticide have been those reported for langur groups, primarily Presbytis entellus (Blaffer Hrdy, 1974; Sugiyama, 1975), but also P. senex (Rudran, 1973), and P. potenziani (Ron Tilson, personal communication, cited by Blaffer Hrdy, 1977). These reports focused on the death or disappearance of infants during periods of social unrest surrounding replacement of adult male(s) within the troop by one or more males from outside the troop. This condition apparently corresponds with what we have seen in captive groups of M. nemestrina and M. fascicularis, i.e., there is a strong tendency for males placed in new groups to traumatize the infants and neonates in those groups. Blaffer Hrdy (1974, 1977) has inferred that all dead or missing infants were killed by the new males, whereas only a small number of male killings were actually observed (relative to the number of deaths and disappearances). As noted above, we have also observed females killing their own infants, usually when the female and neonate were being captured for postnatal examination (it would have been as likely to occur if the infant were being separated for nursery rearing at that time). The female's response in this situation seems almost reflexive and it has been observed in wild-born as well as laboratory-born females. To deal with this problem, examination has been postponed for a few days in hopes that the behavioral pattern will subside. This procedure has apparently succeeded, although it has not been fully evaluated.

As Blaffer Hrdy (1977) has correctly pointed out, the basic issue with regard to infanticide is the extent to which "social pathology" resulting from environmental stresses is responsible for the phenomenon. This is obviously important with regard to mass production of primates in captivity, for if the phenomenon occurs regardless of environmental influences, we cannot hope to improve the situation by manipulation of environmental factors. If, however, we are dealing with pathogenic environments, there is a possibility of altering conditions (e.g., overcrowding) in ways that will eliminate the undesirable behavior. Blaffer Hrdy (1977) suggests that environmental factors are not primarily responsible, and that infanticide is instead a biologically programmed behavior with adaptive consequences for the male perpetrators. If this is true, our attention should probably focus on maintaining stable groups with somewhat less concern over the physical environment.

My own feeling is that this is not a simple matter and that it is essential to consider several factors simultaneously. I have conducted a series of studies at the Primate Field Station that point to the necessity of evaluating social factors and features of the physical environment as they interact with each other.

Infant trauma, like other bite wounding in the colony, follows two basic patterns: acute episodes such as those described above in which the perpetrator is usually a male or the infant's own mother, and chronic cases in which the perpetrator is usually the mother or some other adult female. Adult females housed in groups also sustain a substantial number of bite wounds, the majority apparently resulting from chronic aggressive interactions with other females.

EFFECTS OF CROWDING

Crowding has been implicated as a contributor to pathological patterns of behavior, including hyperaggressivity, infanticide, and cannibalism in rodents (Calhoun, 1961), and a direct link between population density and aggression has generally been assumed (Ardrey, 1966; Lorenz, 1966). For primates (including humans), however, there is little empirical evidence of such a direct relation between crowding and violence.

Southwick (1967) reported that aggressive episodes doubled when rhesus monkeys were confined to one-half their usual enclosure. This result has frequently been cited as evidence of a direct relation between crowding and aggression. There are, however, reasons to regard this interpretation with some caution.

When the space was reduced to one-half that normally available, the opportunity for interaction doubled. Thus, the expected result of increased spatial density based solely on opportunity seemed to be confirmed, but no additional effect attributable to crowding stress was indicated. The crowding was accomplished by placing a fence down the middle of the enclosure, herding the monkeys through a small gate, and closing the gate. Aggressive encounters were counted before and after the barrier was in place and after the gate was closed. The significant difference reported was that between the baseline (no barrier) and the crowded (gate closed) phases. There was no significant difference between the baseline period and the period during which the barrier was in place with the gate open, allowing access to both halves of the enclosure; nor was there a significant difference between the period during which the barrier was in place with the gate open and the period during which the gate was closed. As the most meaningful comparison would have been between the latter two conditions, no crowding effect was actually demonstrated. In fact, aggression increased more when the barrier was placed (relative to the baseline period) than when the animals were crowded into one-half the usual space. In addition, the two halves of the enclosure were not identical, for there was a pool on one side and distribution of tree stumps and other perches was assymetrical. Thus, a number of qualitative changes in the environment coincided with the quantitative spatial change, and an evaluation of purely spatial density factors was not possible. In sum, we must conclude that considerably less aggression occurred under the crowded condition than would have been expected on the basis of increased opportunity for interaction or intensification of crowding stress.

Alexander and Roth (1971) reported that aggressive interactions among male Japanese monkeys (M. fuscata) were more frequent when a troop was confined to a small holding pen than when it was in a large corral. Aggressive interactions were considerably less frequent, however, among females in the crowded condition than in the usual larger area. This result was consistently replicated several times under the same conditions. The qualitative differences between the two conditions were considerably more pronounced in that study than in the one by Southwick (1967). Nevertheless, the sex-specific responses suggest that the size of an enclosure may be more critical for multimale groups than for harem or all-female groups. The speculation that reduced space might actually reduce female-female aggression (an important source of chronic trauma problems) is inescapable. Some supporting evidence from research at the Medical Lake facility is discussed later in this chapter.

Only one nonhuman primate crowding project has been reported in which qualitative aspects of the environment were not confounded with experimental quantitative spatial manipulation. A small harem group of baboons (Papio cynocephalus) was studied for an extended period (Elton, in press). A movable wall was used to reduce available space in several stages until one-half the original space remained. Through successive stages aggressive episodes at first increased and then decreased in frequency. The adult male began to exhibit some abnormal behaviors including aggression toward juveniles under the most crowded conditions. Unfortunately, the project was discontinued before the reverse phase (gradual expansion to the original room size) was instituted. Nevertheless, it provided some rather convincing evidence of pathological behavior associated with extreme crowding!

It seemed plausible that some of the aggressive behavior resulting in trauma at the Medical Lake facility might have been a consequence of overcrowding, so I set out to test this possibility. At the time this series of studies was initiated, harem groups were housed in two-room suites, each room measuring 2.2 x 3.1 x 2.8 m. A small shuttle door (0.5 x 1 m) at floor level allowed access between rooms. Each room contained a perch, covered radiator, ad lib water supply, and a drain. Once each day all the animals in each group were crowded into one room while the other was cleaned. It seemed likely that aggression would intensify during that period if crowding was really a contributing factor.

Methods

The effects of short-term crowding on aggression were observed in 14 groups containing 208 M. nemestrina by counting absolute frequencies of contact aggression (grab, push, hit, bite), chase, threatened aggression (open-mouth threat), harsh vocalization, grimace, screech, and crouch. Observations were made when animals had access to both rooms or only one room of their suites. The first six groups were tested in two three-period sequences with each period lasting 20 minutes. In one sequence the groups were observed first in two rooms, then in one room, and again in two rooms; the other sequence was the opposite. The other eight groups underwent the same two sequences plus two additional ones involving no spatial change, i.e., three consecutive 20-min periods each in one room and two rooms. The age and sex of the perpetrator and recipient of each aggressive act were recorded.

Results

Aggressive interactions involving adult males or infants were infrequent relative to those involving adult females, and no spatial effects on behaviors of males or infants were apparent. Females, however, exhibited considerably more aggression under the less crowded condition than under the crowded condition. Contact aggression (grab, push, hit, bite) was nearly three times as high when two rooms were available (204 episodes) as when the groups had access to a single room (75 episodes; $p < 0.02$, Wilcoxon test). Frequencies of open-mouth threats were consistent with those of contact aggression (188 in one room; 519 in two; $p < 0.02$, Wilcoxon test). For a complete description of the results of this study see Anderson et al. (1977).

Superficially the results seemed to indicate that increased density reduced aggression, but there was a paradoxical aspect: the frequency of contact aggression per capita among females correlated significantly with the number of females per group (rho = 0.62, $p < 0.02$). Thus, frequency of contact aggression was directly related to one dimension of crowding, "social density" (size of group), and inversely related to another dimension, "spatial density" (amount of available space).

Subsequent studies confirmed the relation of group size to frequency of aggression, and contact aggression was even more frequent relative to the number of females per group than would be expected according to a positively accelerating function describing the increased opportunity for interaction as group size increased from 8 to 13 females (Erwin and Erwin, 1976; Erwin, 1977b). Thus, the number of females per group emerged as a major risk factor with very little contact aggression occurring in groups containing 8 or fewer females and very high levels occurring in groups containing more than 10 females.

Additional research also revealed a social factor that was probably responsible for the unusual result of the spatial density manipulation. Among some of the groups surveyed at the Primate Field Station, no male was present. In a few other groups there were two or more males, while most contained only one male. Aggression among females was inordinately high when no male was present, but the presence of an adult male apparently inhibited aggression among females. There was essentially no additional decline in female-female aggression when more than one adult male was present. This result was confirmed by additional surveys and experimental studies (Sackett, Oswald, and Erwin, 1975; Oswald and Erwin, 1976; Dazy, Kuyk, Oswald, Martenson, and

Erwin, 1977). It became clear that there was an interaction between social and spatial factors in the one-room versus two-room study. The male occupied only one of the two rooms at any one time and consequently inhibited aggression only in that room. The females were free to fight in the other room of the suite. In fact, aggressive bouts lasted longer with a higher frequency of bites in the room in which the male was absent. Thus, it was not the amount of space that was important, but the quality of space.

Shortly after completion of this study on short-term crowding, the colony records showed that treatment for trauma was more frequent after adoption of the two-room housing than during a comparable period during which groups had been housed in single rooms. The housing was changed back to single-room and the rate of trauma decreased.

In another study (Erwin, Anderson, Erwin, Lewis, and Flynn, 1976), a concrete pipe was introduced into rooms housing pigtail groups. In groups that remained stable, the rates of aggressive behavior diminished when "cover" was available. Frequently aggressed animals escaped into the concrete pipes, usually ending aggressive encounters. Groups that had substantial changes in social composition did not benefit from this environmental manipulation; in fact, several females were seriously wounded in a group that underwent an exchange of resident males. This study, along with other evidence on the consequences of changes in group composition, resulted in a major change in colony procedures. Special breeding groups were designated for maintenance in complete stability with no exchange of adult males or addition of unfamiliar females. A system was devised for ranking breeding success of groups with regard to pregnancy rates and treatments for trauma or disease, and consistent maintenance of the highest ranking groups was guaranteed.

The supervisory veterinarian of the Primate Field Station recently reported that the number of treatments for trauma during 1976 was about one-sixth that for the year before instituting the changes recommended by these studies. While I do not suggest that the answers to trauma problems are the same for all colonies, I do suggest that there is no substitute for careful observation and measurement of the effects of environmental, social, and developmental parameters on the functioning of domestic breeding colonies of nonhuman primates. This can be effectively accomplished only by keeping systematic records and using scientific methods to determine colony husbandry procedures, including quantified observation of behavior.

ACKNOWLEDGMENTS

The work reported here was conducted at the Regional Primate Research Center Field Station at Medical Lake, Washington, supported by NIH grant RR00166.

REFERENCES

Alexander, B. and Roth, E.   The effects of acute crowding on aggressive behavior of Japanese monkeys.   Behaviour 39:73–90, 1971.

Anderson, B., Erwin, N., Flynn, D., Lewis, L., and Erwin, J.   Effects of short-term crowding on aggression in captive groups of pigtail monkeys (Macaca nemestrina).   Aggres. Behav. 3:33–46, 1977.

Ardrey, R.   The Territorial Imperative, New York: Atheneum, 1966.

Blaffer Hrdy, S.   Male-male competition and infanticide among the langurs (Presbytis entellus) of Abu, Rajasthan.   Folia primat. 22:19–58, 1974.

Blaffer Hrdy, S.   Infanticide as a primate reproductive strategy.   Amer. Sci. 65:40–49, 1977.

Burton, F.   The integration of biology and behavior in the socialization of Macaca sylvana of Gibralter.   In: Primate Socialization, F. Poirier (Ed.), New York: Random House, 1972.

Calhoun, J.   Phenomena associated with population density.   Proc. nat. Acad. Sci., Wash. 47:428–449, 1961.

Carpenter, C. R.   Societies of monkeys and apes.   Biol. Sym. 8:177–204, 1942.

Dazey, J. and Erwin, J.   Infant mortality in Macaca nemestrina: Neonatal and post-neonatal mortality at the Regional Primate Research Center Field Station at University of Washington, 1967–1974.   Theriogenology 5:267–279, 1976.

Dazey, J., Kuyk, K., Oswald, M. Martenson, J., and Erwin, J.   Effects of group composition on agonistic behavior or captive groups of pigtail macaques, Macaca nemestrina.   Amer. J. phys. Anthrop. 46:73–76, 1977.

Drickamer, L.   A ten year summary of reproductive data for free-ranging Macaca mulatta.   Folia primat. 21:61–80, 1974.

Elton, R.   Baboon behavior under crowded conditions.   In: Captivity and Behavior: Primates in Breeding Colonies, Laboratories, and Zoos, J. Erwin, T. Maple, and G. Mitchell (Eds.), New York: Van Nostrand Reinhold, in press.

Erwin, J. Infant mortality in Macaca fascicularis: Neonatal and post-neonatal mortality at the Regional Primate Research Center Field Station, University of Washington, 1967–1976. Theriogenology 357–363, 1977a.

Erwin, J. Factors influencing aggressive behavior and risk of trauma in the pigtail macaque (Macaca nemestrina). Lab. anim. Sci. 27:541–547, 1977b.

Erwin, J., Anderson, B., Erwin, N., Lewis, L., and Flynn, D. Aggression in captive groups of pigtail monkeys: Effects of provision of cover. Percep. mot. Skills 42:319–324, 1976.

Erwin, N. and Erwin, J. Social density and aggression in captive groups of pigtail monkeys (Macaca nemestrina). Appl. anim. Ethol. 2:265–269, 1976.

Hird, D., Henrickson, R., and Hendrickx, A. Infant mortality in Macaca mulatta: Neonatal and post-neonatal mortality at the California Primate Research Center, 1968–1972. J. med. Primat. 4:8–22, 1975.

Lorenz, K. On Aggression, New York: Harcourt, Brace, Javanovich, 1966.

Oswald, M. and Erwin, J. Control of intragroup aggression by male pigtail monkeys (Macaca nemestrina). Nature 262:686–688, 1976.

Rudran, R. Adult male replacement in one-male troops of purple-faced langurs (Presbytis senex senex) and its effects on population structure. Folia primat. 19:166–192, 1973.

Sackett, D. P., Oswald, M., and Erwin, J. Aggression among captive female pigtail monkeys in all-female and harem groups. J. biol. Psychol. 17(2):17–20, 1975.

Southwick, C. An experimental study of intragroup agonistic behavior in rhesus monkeys (Macaca mulatta). Behaviour 28:182–209, 1967.

Struhsaker, T. Infanticide in the redtail monkey (Cercopithecus ascanius schmidti). In: Proceedings of the Sixth International Congress of Primatology, London: Academic Press, in press.

Sugiyama, Y. On the social change of Hanuman langurs (Presbytis entellus) in their natural conditions. Primates 6:381–417, 1965.

Valerio, D. Breeding Macaca mulatta in a laboratory environment. Lab. anim. Handb. 4:223–230, 1969.

CHAPTER 19

THE INTELLECTUAL CONSEQUENCES OF EARLY SOCIAL
RESTRICTION IN RHESUS MONKEYS (Macaca mulatta)

John P. Gluck

Department of Psychology
University of New Mexico
Albuquerque, New Mexico

INTRODUCTION

For over a decade the nonhuman primate research concerned
with the effects of early social isolation on behavioral functioning
contained a seemingly puzzling discrepancy. On the one hand,
rhesus monkeys (Macaca mulatta) reared in environments that
either severely restricted or completely occluded social contact
with conspecifics were found to show easily discriminable deficits
in personal, social, sexual, and maternal behaviors (see Chapter
20). On the other hand, Harlow and coworkers have long denied the
existence of similar intellectual deficits (e.g., Rowland, 1964;
Harlow, Schiltz, and Harlow, 1969; Gluck and Harlow, 1971).

The human primate literature concerned with the topic of
intelligence and environment could not contribute to the resolution
of this issue. Writers such as Jensen (1969) and Herrnstein (1971)
made careful and well publicized arguments that stressed the
genetic components of intelligence, and therefore supposed it to be
robust to environmental insult. In addition, studies that emphasized the importance of environmental factors were rendered
merely suggestive by the inherent methodological confoundings
(Spitz, 1945, 1946; Dennis, 1973).

A clearer and more certain picture has begun to emerge. We
now feel confident in stating that not only do learning differences
exist between differentially reared monkeys, but we are beginning
to understand the basis of the observed differences.

The contents of this chapter will proceed in four parts: definition of terms and assessment paradigms, history, recent research, and conclusions. The first section will set definitions relating to the rearing procedures, and the common techniques and apparatus used in the assessment of intellectual functioning of monkeys. The history section will briefly recount the progression of experiments prior to 1970, while the recent research portion will deal with results accumulated since that time. Hopefully, some meaningful meeting of ideas and data will take place in the conclusion.

DEFINITION OF TERMS

Rearing Conditions

To assess the intellectual consequences of differential early experience, we studied four basic rearing conditions: total social isolation, partial social isolation, mother-peer, and nuclear family environments. Total social isolation (TI) involves separating an infant monkey from its mother no later than several hours after birth, and raising it in an enclosed chamber that eliminates visual and tactual contact with conspecifics. Obviously, the chamber also significantly reduces the ambient level of nonsocial stimulation. Partial isolation (PI) is achieved by separating infants at birth and raising them individually in open cages that allow visual, auditory, and olfactory contact, but not physical contact, with other monkeys. Clearly, the level of nonsocial stimulation also exceeds that of the TI conditions. Mother-peer (MP) rearing involves the group housing of mothers and their infants. Typically, the infants are all quite similar in age. The nuclear family (NF) condition (Harlow, Schiltz, Harlow, and Mohr, 1971) provides the developing monkey access to both adult males and females as well as nonadult individuals at all stages of development. In addition, the physical space is both substantially larger in area than the other conditions, and contains a variety of exercise and play apparatus.

Assessment and Methodology

The vast majority of the experiments concerned with intellectual assessment in nonhuman primates have used very similar tests, presented in similar testing environments. The basic testing environment is the Wisconsin General Test Apparatus (WGTA). The apparatus can be conceptually divided into a subject holding section, problem presentation and response section, and an experimenter preparation and observation section. The subject holding section is a cage constructed to limit the overall motor behavior of subjects. The cage is typically large enough to allow comfortable movement but disallows most acrobatics and extensive exploration. A movable opaque screen separates the holding area from the problem presentation space. While a problem is being prepared, the screen prevents visual and manipulatory contact by the subject. The problem presentation area consists of a tray upon which problems are placed. The stimulus objects are placed on specific positions on this tray over small depressions or "wells." Food rewards such as raisins, grape pieces, candies, etc. are placed in these wells under the "correct" object. The tray normally slides on a track, in or out of the animal's reach, to control response opportunity. The experimenter section is partitioned from the presentation and response space by a one-way mirror that permits observation of the animal response, yet reduces the transmission of inadvertent solution cues by the experimenter.

The types of tasks presented in the WGTA are intended to test discrimination ability, inter-problem transfer, immediate memory, and concept learning. The stimuli used in these tests are constructed of "junk" objects (e.g., small toys, blocks, cans, etc.) attached to a wood base sufficiently large to cover the food wells cut into the formboard. With discrimination tasks, the stimulus objects are likely to differ from one another on the basis of size, shape and color. Two-choice object discriminations are performed by consistently placing a food reward beneath the arbitrarily selected "correct" object and varying its left and right positions randomly. Delayed response tasks are conducted differently. The monkey subject is first faced with two identical objects. The experimenter then attracts the animal's attention by placing a highly preferred food bit under one of the objects. Care is taken to insure that the animal has observed this "baiting" phase. Next, a predetermined amount of time, ranging from a few seconds to potentially several minutes, is allowed to expire before the formboard is pushed within response reach of the animal. An oddity problem involves presenting three stimulus objects. Two of the objects are identical, while the third object differs from them on several dimensions. The task is for the animal to select the "odd" object of the three regardless of its position. This type of problem is quite difficult for monkeys and is seen as requiring the acquisition of a relational concept. A discrimination reversal task begins by establishing a two-choice object discrimination to some predetermined high level of correct performance. Next, the reward contingencies are reversed; that is, responses to the previously rewarded object now go unrewarded, while responses to the previously unrewarded object now lead to the presentation of reward. Finally, inter-problem transfer or learning set (Harlow, 1949) refers to the fact that when subjects are presented many separate exemplars of each of these problem types, performance on the later problems is superior to performance on the initial items. In other words, the animals seem to adopt a method of solution or strategy that applies to the entire class of problems. In Harlow's (1949) words, "the animal learns how to learn."

RECENT HISTORY

In 1962, Mason and Fitzgerald tested a single rhesus monkey that had been raised from birth in TI. They tested their subject on problem types that they believed demonstrated basic intellectual ability. They reported that performance on two-choice discriminations, discrimination reversal, and learning set were all within the range of previously tested, nonisolated monkeys. Guy Rowland dedicated his doctoral dissertation to a detailed examination of the apparent discrepant effects of social isolation on learning performance and social functioning. Rowland (1964) raised four groups of monkeys: TI for the first 6 months of life, TI for the first 12 months of life, TI for the second 6 months of life, and a group of controls raised in PI for the first 12 months of life. Beginning at 45 days of age all subjects were adapted to the WGTA. The adaptation procedure used by the Harlow group is quite comprehensive, and includes a long period of exposure to both the experimenter and the apparatus; slow, graded training in all aspects of response performance; and a series of 45 "practice" problems prior to exposure to the experimental problems. Next, the subjects were tested on 400 6-trial, two-choice object discriminations, and on 500 0- and 5-sec delayed response problems. Finally the subjects were tested for their ability to learn to avoid electric shock. All subjects were placed in a large "shuttle box." A tone signaled a period when movement to the other side of the apparatus would result in the avoidance of shock. However, if a shuttle response was not made while the tone was on, shock was presented, and remained on until the subject escaped to the safe side of the apparatus. All subjects were also tested for social functioning in a variety of social behavior tests. The results of this comprehensive set of experiments revealed that, although major social behavior differences existed between the groups, asymptotic performance on all learning tests was indistinguishable from that of the PI controls. Conceptually, the results of Rowland's experiments solidified the notion that social behavior was susceptible and intellectual activities were immune to early environmental insult.

Griffin (1966) continued the analysis of early experience effects on intellectual and social behavior by rearing five rhesus monkeys in TI for the first 3 months of life. Upon release from the rearing environments, one animal died of starvation and a second required heroic intervention to survive. During subsequent social testing, the four surviving monkeys showed substantial and adequate behavioral adjustment. As expected, tests of two-choice object discrimination learning sets and delayed responses revealed performance in normal ranges.

Harlow and his research group were not the only researchers interested in early experience and primate behavior. Encouraged by the foresight of Henry W. Nissen, studies of the developmental consequences of early social restriction on chimpanzees were begun at the Yerkes Primate Research Center in 1956. Sixteen chimpanzees (Pan troglodytes) were separated from their mothers within 12 hr after birth and were reared in one of several conditions. Total isolation required that the subjects be housed in cribs with barren walls and ceiling. The other rearing conditions added small amounts of visual, manipulatory, and social experience to this baseline condition. Control subjects were jungle-born and captured at approximately one year of age. Following capture, they lived in relatively enriched laboratory conditions. Consistent with the rhesus data, the deprived chimps were found to display extreme fearfulness, bizarre stereotypies, and deficient social functioning (Davenport and Rogers, 1970).

When the subjects were all 7 to 9 years old, tests of intellectual ability were begun. Immediately we encounter an interpretational difference. Harlow has argued that group comparisons should not be made until all subjects have reached an acceptable level of performance (e.g. Harlow et al., 1971). The argument is based on the desire to partition the effects of emotional variables from the effects of "purely" intellectual ones. On the other side of this issue, Davenport and Rogers (1970) stated that it was their purpose to compare multiple behavioral dimensions and they therefore began the experimental comparison at the pre-training stage. They reported that the performance of the wild-born subjects was clearly superior to that of the restricted group during pre-training. Most of the wildborns passed all pre-training requirements during the first day of testing, while the restricted groups acquired them at a much slower pace; indeed, one subject was so slow it had to be dropped from the experiment.

Basically, the restricted chimps engaged in a number of activities (rocking, swaying) that interfered with task-relevant activities.

The chimpanzees were tested on a series of 438 two-choice object discrimination learning set problems. Analysis of the performance revealed that the wildborn controls reached their asymptotic level of performance earlier than the deprived subjects, though no group differences in asymptotic level were discernible by the end of the experiment. The restricted subjects also showed consistent difficulty in problems where their first trial response was incorrect: they seemed unable to break their stimulus preference-based responding (Davenport, Rogers, and Menzel, 1969). Both control and restricted groups were trained and tested on the delayed response task which employed 0- and 5-sec delay intervals (Davenport and Rogers, 1969). Once again, the restricted chimps engaged in many nontask behaviors that interfered with attention to both the baiting and delay phase. However, as experience in the task continued, the restricted group acquired more appropriate behavioral patterns and reached a level of performance equal to that of the controls. It was then reported (Rogers and Davenport, 1971) that restricted chimpanzees showed marked difficulty with changing response strategies when tested on 1,700 trials of oddity problems.

Thus, the work with chimpanzees both supported the contention that emotionality was a likely confounding factor in early experience assessment and presented serious evidence of intellectual disruption that contrasted sharply with Harlow's notion.

One of the most dangerous theoretical positions to find oneself in is that of defending the null hypothesis. Harlow was quite aware of his position in this regard. At one point he even registered surprise at the consistency of the intellectual results and suggested that his small sample sizes might lack the power to uncover subtle learning differences (Harlow, Dodsworth, and Harlow, 1965). Therefore, Harlow adopted the only reasonable experimental strategy available. Specifically, he attacked the intellectual immunity hypothesis on two fronts: he began to include more difficult and complex learning tests, and increased the social sophistication of the control groups.

In 1969 Harlow et al. reported the results of an experiment relevant to the first part of the experimental strategy. Consistent with the previous experiments, monkeys reared for 6 and 9 months in TI were found to perform at levels that equaled the PI controls on object quality learning set and delayed response. In addition, no significant group differences were found on the difficult and demanding set of 256 oddity problems utilized for the first time. In fact, the isolated subjects showed consistent, but not significant, superior performance on the two-choice discriminations.

RECENT RESEARCH

The first crack in the intellectual immunity hypothesis appeared in 1971, when Harlow et al. presented preliminary data from an experiment that compared the performances of a group of PI's and enriched monkeys reared in the NF. In line with earlier work, no significant differences were observed on two-choice discriminations and delayed response tasks. However, a suggestion of difference was offered on the oddity problems. Since the results were based on only a very small number of subjects, strong statements were avoided. Then in 1973, Gluck, Harlow and Schiltz published the results of a comprehensive comparison of PI's and NF controls. Specifically, 12 PI's and 12 controls were tested on 600 two-choice object quality discriminations, 1,920 trials of multiple delayed response, and 256 separate oddity problems. Figure 1 displays the results of the object-quality discriminations. At no

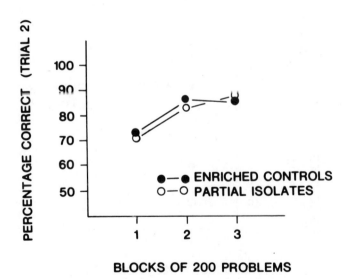

Figure 1.   Percent correct responding on the second trial of 600 two-choice object discriminations.

point does the performance of group NF exceed that of group PI, though both show the expected increase in performance across problem blocks. Similarly, Figure 2 shows that testing and delayed response problems with delayed lengths that varied from 5 to 40 sec fail to differentiate the groups. However, Figure 3 reveals

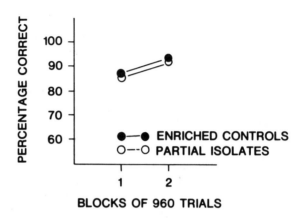

Figure 2.   Percent correct responding on 1,720 trials of multiple delayed response (0-40 sec).

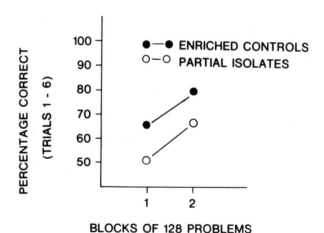

Figure 3.   Percent correct responding during 256 six-trial oddity discriminations.

remarkable differences between the groups during oddity testing. By the second block of 128 problems, the enriched monkeys had stabilized at a performance level of approximately 80%, while the PI's had barely attained an accuracy grade of 65% on trials 1 through 6. As the authors stated, the results of this experiment forced a significant change in stance on the issue of early restriction and later learning.

Up until this time we were willing to ascribe the chimpanzee results to emotional and adaptation confounding, or else species differences.   However, with the experiments of Gluck et al. (1973), we were completely confident in the effectiveness of the long, tedious pre-experimental adaptation procedure, and therefore resisted invoking those factors in our interpretation. We searched our experience for an explanation. Observation of the monkeys' behavior during the oddity experiment gave us the impression that the PI's were not consistently scanning the middle object in the display. Without consistent use of this information, performance would obviously be depressed.   The question became, if this observation was correct, what would have led to such a scanning strategy, and why was it maintained in the isolates?

One element of the oddity procedure was that the animals were not required to respond to the center position.   That is, the odd object was placed in either the extreme left or right positions at all times. In fact, animals were required to observe and respond to the center position of the formboard only once early during adaptation. Therefore, the animals may have learned to ignore this important source of information. Specifically, after the subjects had become accustomed to the holding section of the WGTA, food was presented in an uncovered center-positioned food well.   The next phase required the monkey to displace a single object that covered the food well.   After that point, the animal was not required to either respond or scan the center position of the formboard for any experimental purpose again for nearly 8,000 experimental trials spread over nearly 2 years. Given the finding that monkeys tend to restrict their observation to response-relevant areas (Meyer, Treichler, and Meyer, 1965), it is not difficult to imagine how a scanning strategy with a center-position neglect might be established. Why would this strategy continue to be exhibited by the isolate subjects despite the reward cost?

After a complete review of the early experimental and primate social developmental literature, Sackett (1970) proposed a theory of the effects of early social isolation. According to this proposal, restriction is particularly damaging to neural mechanisms concerned with the cessation of responding. In other words, once a response has become highly probable, isolated monkeys have specific difficulty in inhibiting this response in the face of changed environmental conditions. Davenport, Rogers and Rumbaugh (1973) offered experimental support for this position. In their experiment, chimpanzees reared in either jungle or TI environments were trained to specific accuracy criteria on two-choice object discriminations. Once achieved, the reinforcement conditions on each problem were reversed. Results clearly demonstrated that the isolates persisted with the previously correct, but now inappropriate, object choices, while the feral-reared controls did not.

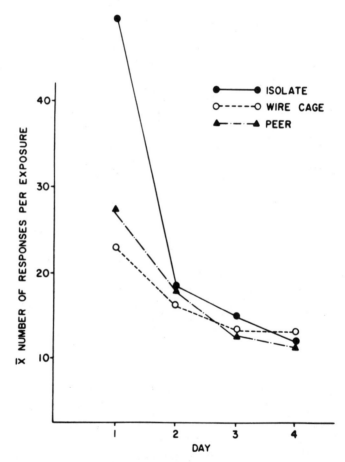

Figure 4. Number of unrewarded level press responses emitted following initial response acquisition.

Since all subjects were trained to identical initial acquisition criteria, explanations that invoke differential adaptation confoundings are ruled out.

Encouraged by these findings, Gluck and Sackett (1976) trained three groups of differentially reared monkeys (TI, PI, MP) to press a level for sucrose reinforcement. Each lever response was rewarded with a 25-mg sucrose tablet. All subjects were given equal numbers of rewarded lever presses. Next, rewards were no longer delivered following a lever press response. It was found that TI's emitted many more nonrewarded lever press responses than either of the two remaining groups, as shown in Figure 4. Parenthetically, similar evidence of unusual response persistence has been reported for rats reared in social isolation (e.g., Gluck and Pearce, 1977).

Even more recently, Frank, Gluck, and Strongin (1978) have extended their findings of response persistence. In their experiment, TI's and MP's were trained to lever press for food reinforcement. When both groups were responding at an even and steady response rate, a 1,400-Hz tone (conditioned stimulus) was presented several times per session for 3 min. At the conclusion of each stimulus presentation subjects received an unavoidable electric foot shock (unconditioned stimulus). The number of lever press responses emitted during the 3-min stimulus presentation period prior to the shock was compared with stimulus-free periods. As expected, the MP group learned to suppress responding during the presentation of the shock-predicting stimulus. Surprisingly, yet consistent with the previous data, restricted monkeys persisted in responding during these times (see Figure 5). Behavioral observation of these monkeys showed that they "expected" the upcoming shock, but continued to respond nonetheless.

This set of experimental results is fairly explicit in its interpretation: Monkeys and chimpanzees reared in restricted environments have difficulty altering high probability responses made inappropriate by changed environmental demands. Therefore, the "cognitive" and "intellectual" differences may result from a fairly limited, but highly influential, altered response mechanism.

CONCLUSION

The preceding experimental scenario leaves little doubt that restricted monkeys and apes are cognitively inferior to socialized controls on tasks which require the cessation of high probability responses. Evidence (Gluck and Sackett, 1976) further suggests that the severity of this deficit is correlated with severity of the restriction. Taken together, the evidence supports a theory of early experiment effects offered by Sackett (1970). This theory postulates that the absence of variable stimulus input and feedback during development insults the neural mechanisms concerned with response inhibition. Although Sackett does not speculate on the specific mechanism involved, there are some exciting hints in the literature. For example, Douglas (1972) presents convincing evidence that the hippocampus is the structure primarily responsible for the cessation of responding due to nonreinforcement. The experimental effects of hippocampectomy are strikingly similar to

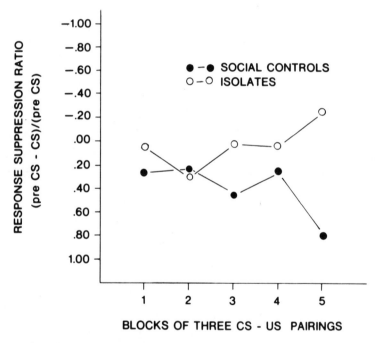

Figure 5.   Amount of lever press response suppression seen in response to the shock predicting stimulus. The ratio is calculated by subtracting the number of responses made during a sample of stimulus-free periods from stimulus-present periods and dividing this amount by the number of responses made during the stimulus-free periods. Negative numbers reflect no suppression; positive numbers reflect disruption.

the data we have presented in this chapter. Specifically, hippo-campectomized rats and monkeys show retarded extinction of previously reinforced responses and slower rates of performance during discrimination reversal; and they require a longer habituation time, and are more difficult to distract while engaged in well learned goal-directed activity (Douglas and Pribram, 1966; Douglas, 1967; Kimble, 1968; Douglas and Pribram, 1969).

It is, of course, conceivable that the behavioral differences do not reflect neurological changes. It may well be that behavioral habits learned in the restricted environment are simply revealed in the testing situations. Perhaps living in an environment in which the bulk of stimulus changes do not predict or accompany contingency change instructs the inhabitant to ignore the occurrence. In other words, restriction-reared subjects have not learned to identify or assess environmental change; in fact, they have learned the opposite. There is, however, experimental evidence that seriously conflicts with this line of reasoning. It has been reported that although restricted chimpanzees housed in mixed social groups have begun to acquire an appropriate repertoire of species-typical social responses, the cognitive deficits clearly continue to emerge during testing (Davenport and Rogers, 1970).

ACKNOWLEDGMENTS

This chapter is dedicated to the memory of Kenneth A. Schiltz, who was to a great degree the Wisconsin research tradition. Kenny taught so many of us guidelines that led first to our survival when we encountered our first monkey, and eventually to our success as researchers. I remember when H. Harlow first directed me to his office. He sarcastically said that Ken was the second best person in the world with the WGTA. I was, of course, to discover that Ken was indeed the best.

REFERENCES

Davenport, R. K. and Rogers, C. M. Intellectual performance of differentially reared chimpanzees: I. Delayed response. Amer. J. ment. Defic. 73:963-969, 1969.
Davenport, R. K., Rogers, C. M., and Menzel, E. W. Intellectual performance of differentially reared chimpanzees: II. Discrimination learning set. Amer. J. ment. Defic. 73:963-969, 1969.

Davenport, R. K. and Rogers, C. M. Differential rearing of the chimpanzee. Chimpanzee 3:337-360, 1970.

Davenport, R. K., Rogers, C. M., and Rumbaugh, D. M. Long-term cognitive deficits in chimpanzees associated with early impoverished rearing. Develop. Psychol. 9:343-347, 1973.

Dennis, W. Children of the Creche, New York: Appleton-Century-Crofts, 1973.

Douglas, R. J. The hippocampus and behavior. Psychol. Bull. 67:416-442, 1967.

Doulgas, R. J. and Pribram, K. H. Learning and limbic lesions. Neuropsychologica 4:197-220, 1966.

Douglas, R. J. and Pribram, K. H. Distraction and habituation in monkeys with limbic lesions. J. comp. physiol. Psychol. 69:473-480, 1969.

Douglas, R. J. Pavlovian conditioning and the brain. Pp. 529-553 in: Inhibition and Learning, R. A. Boakes and M. S. Halliday (Eds.), New York: Academic Press, 1972.

Frank, R. G., Gluck, J. P., and Strongin, T. S. Response suppression to a shock-predicting stimulus in differentially reared monkeys (Macaca mulatta). Develop. Psychol. 13:295-296, 1977.

Gluck, J. P. and Harlow, H. F. The effects of deprived and enriched rearing conditions on later learning: A review. Pp. 103-119 in: Cognitive Processes of Nonhuman Primates, L. E. Jarrard (Ed.), New York: Academic Press, 1971.

Gluck, J. P., Harlow, H. F., and Schiltz, K. A. Differential effect of early enrichment and deprivation on learning in the rhesus monkey (Macaca mulatta). J. comp. physiol. Psychol. 84: 598-604, 1973.

Gluck, J. P. and Sackett, G. P. Extinction deficits in socially isolated rhesus monkeys (Macaca mulatta). Develop. Psychol. 12:173-174:1976.

Gluck, J. P. and Pearce, H. P. Acquisition and extinction of an operant response in differentially reared rats. Develop. Psychobiol. 10:143-149, 1977.

Griffin, G. and Harlow, H. F. Effect of three months of total social deprivation on social adjustment and learning in the rhesus monkey. Child Develop. 37:533-547, 1966.

Harlow, H. F. The formation of learning sets. Psychol. Rev. 56:51-62, 1949.

Harlow, H. F., Schiltz, K. A., and Harlow, M. K. Effects of social isolation on learning performance of rhesus monkeys. Pp. 178-185 in: Proceedings of the Second International Congress of Primatology, Vol. 1, Basel/New York: Karger, 1969.

Harlow, H. F., Dodsworth, R. O., and Harlow, M. K. Total social isolation in monkeys. Proc. nat. Acad. Sci., 54:90-97, 1965.

Harlow, H. F., Schiltz, K. A., Harlow, M. K., and Mohr, D. J. The effects of early adverse and enriched environments on the learning ability of rhesus monkeys. Pp. 121-148 in: Cognitive Processes of Nonhuman Primates, L. E. Jarrard (Ed.), New York: Academic Press, 1971.

Herrnstein, R. J. IQ in the Meritocracy. Boston: Little, Brown and Co., 1971.

Jensen, A. R. How much can we boost IQ and scholastic achievement. Harvard Educ. Rev. 39:1-122, 1969.

Kimble, D. P. The hippocampus and internal inhibitions. Psychol. Bull. 70:285-295, 1968.

Mason, W. A. and Fitzgerald, F. C. Intellectual performance of an isolation reared monkey. Percept. mot. Skills 15:594-596, 1962.

Myer, D. R., Treichler, F. R., and Meyer, P. M. Discrete trial training techniques and stimulus variables. Pp. 1-45 in: Behavior of Nonhuman Primates, A. Schrier, H. Harlow, and F. Stollnitz (Eds.), New York: Academic Press, 1965.

Rogers, C. M. and Davenport, R. K. Intellectual performance of differentially reared chimpanzees: III. Amer. J. ment. Defic. 75:526-530, 1971.

Rowland, G. L. The effects of total social isolation upon learning and social behavior in rhesus monkeys. Unpublished doctoral dissertation, University of Wisconsin, 1964.

Sackett, G. A. Innate mechanisms, rearing conditions, and a theory of early experience effects in primates. Pp. 11-53 in: Miami Symposium on the Prediction of Behavior: Early Experience, M. R. Jones (Ed.), Miami: University of Miami Press, 1970.

Spitz, R. A. Hospitalism: An inquiry into the genesis of psychiatric conditions of early childhood. Psychoanal. Stud. Child 1:53-74, 1945.

Spitz, R. A. Anaclitic depression: An inquiry into the genesis of psychiatric conditions of early childhood. Psychoanal. Stud. Child 3:313-342, 1946.

CHAPTER 20

EXPERIMENTAL AND HUSBANDRY PROCEDURES: THEIR IMPACT ON DEVELOPMENT

G. C. Ruppenthal and G. P. Sackett

Child Development and Mental Retardation Center
and Regional Primate Research Center
University of Washington
Seattle, Washington

INTRODUCTION

This chapter will review a number of studies concerning the effects of varied rearing experiences on the development of monkeys born both in the wild and in colony or laboratory environments. Studies have been chosen which have implications for using monkeys produced in captive colonies as subjects in biomedical and behavioral research. Although common sense tells us that experiences during infancy must be important for normal development in primates, actual practice in nonhuman primate colony stiuations often fails to reflect common sense in toto with respect to the social and environmental stimulation afforded growing infants. This chapter will briefly discuss effects of social deprivation rearing, but more importantly, will describe some less well-known but potentially important effects of variations in specific types of social experience among infant monkeys.

SOCIAL EXPERIENCE

Rearing monkeys in social isolation from other members of the species has devastating effects on behavioral development (Harlow and Harlow, 1965; Sackett, 1972). Total social isolation environments include any situation in which the infant primate can neither see nor physically interact with other members of the species. Some behaviors affected by this extreme form of deprivation include eating and drinking habits, play, exploration, reactions to painful stimulation, maternal motivation and skill, aggression, and reproduction. Two prominent, bizarre, abnormal behavior patterns often displayed by total social isolates are autistic-like self-clutching withdrawal reactions and self-directed aggression. Most of these debilitating effects have been resistant to any type of therapy applied after 6-12 months of age. In one study, however, isolate behaviors were somewhat ameliorated by behavioral therapy in rhesus macaques: success occurred when 6-month-old total isolates were housed with persistent, clinging, 3-month-old normal infants (Soumi and Harlow, 1972).

Monkeys raised in partial isolation (housed in bare wire cages where they can see, hear, and smell--but not touch--other animals) also show abnormal behaviors. These include low levels of social interaction as older infants, juveniles and adults; stereotyped locomotor activities; withdrawal from novel or complex environments; deficits in maternal and sexual behavior; and inappropriate high levels of aggression (Sackett, 1972).

Total and partial social isolation during infancy also produce physiological effects, including abnormal cortisol reactions to environmental stress (Sackett, Bowman, Meyer, Tripp, and Grady, 1973); abnormal reactions to painful stimulation (Lichstein and Sackett, 1971); and hyperphagia and polydypsia (Miller, Caul, and Mirsky, 1971).

Researchers using animals that were raised in total or partial isolation should therefore understand that they may be dealing with deviant individuals yielding research data that may be inappropriate for the problem being studied. Colony managers who are rearing monkeys under such conditions should expect to have few colony-born adults capable of the reproductive and maternal behaviors necessary for maintaining a self-perpetuating breeding colony.

URBAN VERSUS RURAL MONKEYS

Singh (1969) conducted a number of important studies showing that the habitat of feral rhesus monkeys living in India has important effects on adult behavior patterns. He captured monkeys in urban city environments and in rural forested areas, brought them into a laboratory situation, and observed them in a variety of situations. On tests of exploratory behavior, urban monkeys were more likely to approach and interact with novel and complex visual and auditory stimuli than were rural subjects. On social behavior tests, rural monkeys engaged in more affiliative behaviors such as grooming and passive physical contact. Urban monkeys had much higher levels of aggression and threat, especially when paired with forest-captured strangers. However, no differences were found on tests of learning ability.

Thus, the personality of wild-born animals appears to vary with rearing habitat. Rural monkeys appear to be more "friendly," passive, noncurious, and socially subordinate when grouped with urban animals. This study suggests that aggression levels will be high if animals trapped from different habitats are grouped in colony situations. Thus, whenever possible, colony managers should know the trapping site of wild-captured monkeys in order to group animals together who have maximally similar "personalities." To this end, some institutions now are sending on-site veterinary teams to enhance survival and reduce losses of animals at the capture areas.

This principle is also supported by a study measuring preferences of laboratory-raised monkeys for like-reared versus unlikereared animals (Pratt and Sackett, 1967). Two-year-old monkeys that had been reared in total isolation, partial isolation, or with other infants during the first 9 months of life were given a choice between monkeys reared as they were or ones reared in the other two conditions. On some trials the stimulus monkeys were familiar to the subject while on others they were complete strangers. The results clearly showed that like-reared animals preferred each other, even when they were strangers. This preference for likereared animals can also be found in adult, laboratory-raised subjects that are paired together after spending years living in wire cages as juveniles and young adults (Sackett and Ruppenthal, 1973a). Colony managers could take advantage of these findings by grouping together monkeys of like history. This might pay off in reduced social stress and lower rates of aggression with concomitant lowered incidence of disease and mortality, as well as a possible increase in live-born offspring that survive the infancy period.

MATERNAL EFFECTS

The importance of having a "good" mother has long been accepted by child psychologists as a major factor in development of normal social behavior. Many researchers and primate colony managers seem to assume that the sheer presence of a mother will lead primate infants to develop species-typical behaviors. Recent studies concerned with the complexity of the nonsocial environment, peer interaction afforded a mother-raised infant, prior social experience of the mother, age of weaning, maternal parity and age, and social structure of groups containing mother-infant pairs have challenged the simplistic idea that social and emotional development is simply a matter of having good maternal experiences as an infant.

This research trend was begun by the pioneering work of Harlow (1958), who raised rhesus newborns on various types of "surrogate" mothers. Harlow found that "monkey mother love" did not come from being fed by a nurturing mother. Rather, attachment to a surrogate mother was generated only when the mother provided a soft, comfortable, warm surface to which the infant could cling. Such a mother also provided emotional security when the infant was faced with novel or potentially dangerous stimulation. However, even though a surrogate mother could provide the caregiving and security that the infant needed, later studies showed that such surrogate-raised monkeys were grossly abnormal when tested as juveniles and adults. This work has been reviewed by Sackett and Ruppenthal (1973a). Thus, early experience with a permissive, nurturing "mother" does not appear to be a sufficient condition for the development of species-typical behaviors. The remainder of this section summarizes studies that bear on the issue of maternal effects and behavior development of monkeys.

Mother-Only Rearing

Alexander (1966) reared rhesus infants with their mothers for the first 8 months of life. A control group had daily interaction with peers throughout infancy. Two experimental groups were deprived of peer experience for 4 or 8 months from birth. At the end of the rearing period all groups were tested for social behavior with agemates. The only major behavioral difference observed during tests conducted between 8 and 10 months of age was that the mother-only group deprived of peer contact for the first 8 months of life was more aggressive toward familiar and unfamiliar agemates on initial contact than were the other two groups. However, these heightened aggression levels seemed to be transient. This led to the conclusion that the mother herself could provide sufficient socialization for the development of normal social interactions, even in the absence of peer contact.

At 3 years of age these subjects were housed together in pens (Ruppenthal, Sackett and Dodsworth, unpublished observations). A male and a female from each condition lived together in groups of six. After 6 months of pen housing the monkeys were grouped together in a large compound at the Madison, Wisconsin zoo. Finally, the animals were returned to the pen-housing situation in the laboratory. A number of dominance tests, using a water competition procedure, were conducted both in the pens and at the zoo. In this test the subjects were deprived of water for 24 hr. Time taken by each animal to attain 30 sec of drinking was used to measure competitive dominance rank. The animals who had mother and peer experience from birth were most dominant on these tests. This was also reflected in their threat and physical aggression scores taken during periodic social behavior measurement in the pens and at the zoo. Animals deprived of peer interaction for the first 8 months of life were intermediate in threat-aggression levels. The 4-month peer-deprived monkeys were uniformly subordinate throughout the 1½ years of testing.

This study shows that both mother and peer experience are necessary for colony-born monkeys to attain high dominance ranks as juveniles and young adults. Furthermore, effects of particular rearing conditions may not be apparent in tests conducted immediately after termination of infancy experiences. Rather, such effects may only appear later in life, when particular behaviors such as dominance displays and aggression have reached a mature level. Finally, it appears that the relation between mothering and age of introduction to peers during infancy is not a simple one. More research is clearly necessary to determine exactly what types of groupings and what ages of introduction to agemates are necessary for producing relatively dominant or relatively subordinate adults.

Cage Size and Complexity

A few studies of pigtail macaque mother-infant pairs have assessed the effects of cage size and opportunity for nonsocial exploration and manipulation experiences on infant development. Castell and Wilson (1971) reared mother-infant pairs in small or large cages, comparing these experimental groups with pairs that lived together in a large group. Group-reared pairs spent more time with each other during the first 2 months, with the initial excursions from the mother occurring later for group-reared than for singly-caged pairs. During months 5-6, group-living pairs separated from each other more frequently than mother-infant pairs living in either small or large cages. Punishment of the infant rarely occurred in the group situation, but occurred very frequently in the small single cages. In another study (Jensen, Bobbitt and Gordon, 1969) complexity of the living cage was directly related to the amount and consequences of punishment that mothers directed toward their infants. In a stimulus-enriched environment, punishment was more effective in initiating breaks in contact and promoting other activities by the infant such as exploration. In a stimulus-impoverished environment, punishment was more frequent but much less effective in redirecting the infant's behaviors away from the mother. The impact of early deprivation caused by small cage size and lack of stimulation is seen in the results of dominance testing of adolescent pigtail macaques (Jensen, Bobbitt and Gordon, 1971). Animals from deprivation environments were almost uniformly subordinate when paired with animals reared in enriched, larger cage environments. In the only reversal observed, a 2-year-old animal reared in deprivation was dominant over a 1-year-old animal from an enriched environment.

Abusive Mothers

Mitchell, Arling and Møller (1967) studied the behavior of 32 rhesus adolescents that had been reared with their mothers for the first 8 months of life in a playpen apparatus that allowed peer contact on an experimenter-controlled schedule (apparatus described by Hanson, 1966). Of the 32 mothers, 28 were jungle-reared and 4 were lab-reared. The experiment was conducted when their progeny were between 15 and 39 months of age. Average postrearing age was 20 months. The subjects were divided into either High Punishment or Low Punishment groups by ranking them according to the total amounts of brutality, punitiveness, or rejection received from their mothers during the first 3 months of life. Animals from the High Punishment group were found to be more aggressive toward adult and neonate stimuli than were animals from the Low Punishment group. After the experiment had ended, several of the animals were group-housed for brief periods, during which time the two most frequently punished males from the High Punishment group severely injured one female cagemate and killed a second female. Sackett (1967) observed the aggressive behavior exhibited by these same animals after they had neared or reached sexual maturity and found inordinately high aggression levels persisting even at that age. These studies suggest that experiences with an abusive mother during even the first few months of life can have lasting and important effects on offspring behavior. Thus, it is expected that such animals will disrupt normal group formation as both juveniles and adults because of their heightened and persistent aggression when they are grouped with monkeys from different infancy backgrounds.

Multiple Separations

Disrupting mother-infant relationships for procedures such as weight recording, culture specimens, and blood drawing is part of the regular regimen in many laboratories. If the mother accepts her infant back and the infant accepts the mother's care, these acute separations are thought to be innocuous. There is evidence, however, that such disruption may have lasting effects on the infant. Griffin (1966) studied the effects of single- versus multiple-mothering on development of rhesus infants. Group 1 (normal control) consisted of four mother-infant pairs that were never separated during the first 8 months of the infant's life. Group 2 (separation control) contained four infants that were separated from their mothers for 2 hr every 2 weeks, then returned to their mothers. Group 3 (multiple-mothered) contained four infants that were rotated between four mothers, receiving a new mother after a 2-hr separation every 2 weeks. At 9-10 months of age, infants in the separation control group were socially subordinate to all other monkeys. Thus, repeated separations from their "natural" mother relegated these infants to low levels of social dominance. When tested one year after the final separation, the separation control subjects still exhibited high levels of distress and acted more disturbed than the normal controls or multiple-mothered animals (Mitchell, Harlow, Griffin and Møller, 1967).

When these animals were 2½ years old they were grouped together in pens (Ruppenthal et al., unpublished observations). Each group contained a male and a female from each of the three rearing conditions. Measures of physical aggression and water competition dominance ranking were taken over a 6-month period. The multiple-mothered monkeys were more aggressive and more dominant than subjects from the other two groups. The normal, nonseparated controls were intermediate on these measures, while the separation controls were lowest in both aggression and social dominance. This study suggests that inconsistent mothering and repeated separation from a single mother can have persistent effects on important dimensions of later social behavior. Inconsistent mothering seems to produce aggressive, dominant animals, while repeated separation from one mother appears to have the opposite effect.

Maternal Parity

Another maternal factor that could influence adaptability of offspring is parity. Seay (1966) studied rhesus infants who were reared by either primiparous or multiparous mothers, with peer experiences commencing daily from day 16 after birth. At 6 months of age only minor differences were apparent in the social and individual behaviors of the two groups, leading Seay to conclude that primiparous mothers were effective in rearing their offspring.

All infants from this study were then housed in individual cages until they were 2 years old. At this time we were asked to demonstrate normal social interactions to a group of new graduate students (Ruppenthal, unpublished observations). We grouped the primiparous-mothered juveniles in a playroom and observed an immediate bout of violent, nonstop aggression for the next 15 minutes, at which time we were forced to intervene because of injuries. We then introduced the group of multiparous-mothered juveniles into the playroom and observed high levels of play. It was as if these monkeys had been together daily over the preceding 1½ years. No aggression was seen. This led to a formal study in which primiparous- and multiparous-mothered juveniles were paired with adults, other juveniles, and younger monkeys (Mitchell, Ruppenthal, Raymond and Harlow, 1966). The primiparous-mothered group showed high levels of stereotyped pacing and vocalizations, which were not seen in the multiparous-reared monkeys. The most striking finding was the vastly inferior play by the primiparous-mothered group. Thus, multiparous-mothering appeared to produce a much more "normal" juvenile.

Other evidence indicating effects of parity in the behavior of wild-captured monkeys comes from social stimulus preference studies by Sackett, Griffin, Pratt, Joslyn and Ruppenthal (1967). Nulliparous, primiparous, and multiparous wild-born adult female rhesus monkeys were presented with a choice between a neonate, a 1-year old, a 3-year old, and an unfamiliar adult. All stimuli were females. Nulliparous females showed no preferences among these four stimuli. Primiparous females showed a small preference for the neonate, while multiparous females displayed an overwhelming preference for the neonate stimulus animal.

These studies and observations suggest that maternal parity may have a direct effect on offspring development in monkeys. It is clear that much more work should be done investigating the effects of this variable. However, the evidence to date would suggest that primiparous mothers may produce offspring whose value to a colony as normal breeders may be different from that of offspring of multiparous mothers.

Motherless Mothers

In most cases, rhesus females reared without mothers or peer experience during infancy are grossly inadequate with their first-born offspring (Arling and Harlow, 1967). These "motherless mothers" have been characterized as either 1) indifferent, not physically harming their infant but not providing sufficient care for the infant to survive without human intervention; or 2) abusive, often reacting violently to the infant's touch and at times injuring or killing the infant.

Table 1 shows the percentage of adequate, indifferent, and abusive mothers who had male and female firstborn offspring in one survey of 42 motherless mothers. Adequate care of the infant occurred more frequently for female offspring. No sex difference was noted for indifferent care. However, abuse occurred much more often if the infant was a male (Ruppenthal, Arling, Harlow, Sackett, and Suomi, 1976).

Several of these motherless mothers produced more than one infant. Of 14 who spent two or more days with their first infant and were inadequate or abusive mothers, 12 were adequate with their second offspring. Ten mothers were inadequate or abusive with their first infant and spent less than 24 hr with it. Only one of these was adequate with the second offspring. These data suggest that experience with the first infant, even for abnormal mothers, has a beneficial effect on later maternal behavior. The implication is that inadequate monkey mothers should be given

TABLE 1. SEX OF INFANTS AND MATERNAL RATING

Sex of Infants	Maternal Rating			
	Adequate	Indifferent	Abusive	Total
Male	3 (12%)	10 (42%)	11 (46%)	24 (100%)
Female	7 (39%)	9 (50%)	2 (11%)	18 (100%)
Total	10	19	13	42

experience with an infant (perhaps even at the cost of the future use of that infant) as therapy for later successful offspring rearing behavior. A second implication is that the abnormal maternal care often seen in monkeys and apes at zoos is most likely not the product of inadequate diet or hormone imbalances. Rather, such behaviors probably result from rearing procedures that are incompatible with development of normal socialization during infancy.

Maturity

Rhesus and pigtail macaques can produce offspring by as early as 3-4 years of age. However, full physical maturity is not reached until 6-7 years of age. In most of the laboratory studies of maternal behavior by colony-born monkeys, young adults between 4 and 6 years of age have served as subjects. The studies below suggest that rearing variables may have different impact depending on the age at which adult animals are tested.

Reproductively mature rhesus females ranging in age from 4 to 10 years were tested for social preferences (Sackett and Ruppenthal, 1973b). Stimulus choices included an adult female stranger and a neonate under 30 days of age. Each subject was tested twice, with a 14-month interval between tests. As shown in Figure 1, females under 8 years of age preferred the adult stranger on both of their tests. Females given the first test at 7 years of age preferred the adult, but on the second test they preferred the neonate. Females receiving their first test when they were 8 years old preferred the neonate, and showed no preference when they were 9 years old. Females over 8 years old on either the first or second test showed no statistically reliable preference. These data show that maternal motivation, as indexed by relative preference for a neonate over an adult, changes dramatically at 8 years of age. Thus, it might be expected that high-risk mothers, such as the "motherless mothers" described in the previous section of this chapter, would exhibit less abnormal maternal reactions if they were 8 years or older than if they were under 8 years old. This expectation is at least partially confirmed by information concerning the mean age of primiparous motherless mothers. Those that were adequate in caring for their infants had a mean age of 92 months, those that were indifferent had a mean age of 82 months, while abusive females had a mean age of 76 months on delivery of their offspring ($p < 0.05$).

A second source of evidence showing age-related effects on maternal responding comes in a study assessing aggression toward infants by 8- and 4-year-old rhesus females that had been raised during infancy in partial isolation (Arling, Ruppenthal, and Mitchell, 1969). The 8-year-old partial isolates directed no more aggression toward an infant than did multiparous wild-born rhesus females. Four-year-old partial isolates displayed high levels of

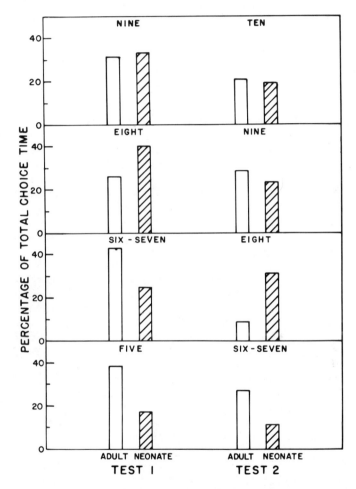

Figure 1. Preference by partial isolate and surrogate-reared females, tested at one-year intervals, for an adult agemate versus a neonate. Subjects within panels are the same individuals on tests 1 and 2. Ages (in years) are given above the bars in each panel.

aggression toward infants. The implication of these studies seems to be that laboratories and colonies that attempt to use socially restricted females as breeders should wait until the animals are mature adults to minimize infant losses and trauma, and to maximize the production of adaptable offspring by such breeders. The years lost in production prior to 8 years of age account for approximately one-third of the animal's reproductive lifetime. Data on rhesus monkeys at the University of Wisconsin Primate Laboratory indicate that females from that colony remain fertile until 20-25 years of age (Mr. Chris Ripp, Jr., personal communication).

EFFECTS OF PROXIMAL HOUSING

There is little sound data available concerning which animals should be paired in forming new social groups under colony or laboratory conditions. Often, violent aggression and death may mark the initial encounters between animals put together into new groups. Our experience with four groups of laboratory-reared rhesus macaques may have some impact in determining group formation. Two of the groups were composed of monkeys that had lived individually in adjacent cages for at leat 6 months before group formation. Members of the other two groups had also lived in individual cages, but these animals were strangers to each other. Aggression levels were much higher in the two groups formed from animals that had lived in adjacent cages. Although the groups were broken up before any deaths occurred, virtually all animals from the familiar groups were treated for lacerations and deep skin bruises caused by bite wounds (Ruppenthal, unpublished observations). We speculated that the neighbor groups responded with aggression because they had all been able to exhibit dominance displays to one another without repercussions while in the individual cage situation. When these animals could physically contact each other, stability of hierarchy was far more difficult, if not impossible, to achieve. It is clear that much more research must be done in this area of "familiarity and contempt" before we can have a rational basis for grouping colony monkeys.

REARING CONDITIONS AND PHYSIOLOGICAL FUNCTIONING

In many laboratories animals are often chosen for invasive and/or terminal medical experiments on the basis of their "expendability." These are often monkeys that are most aberrant in behavior, the poorest for breeding, or somehow undesirable for some other important traits. Often such deviances may be traced to inadequate rearing experiences. The studies in this section suggest that such subject selection procedures may yield invalid data in some cases.

Miller et al. (1971) compared the eating and drinking behaviors of feral adult rhesus macaques with those of monkeys raised in isolation in the laboratory. Before the study started, all subjects were on the once-per-day feeding regimen typical of many laboratories, and at the start of the study the two groups did not differ in body weight. Each individually caged subject was introduced to an ad lib food and water diet for several months. During this time the isolates drank three times as much water and consumed over 50% more calories per day than did the feral animals. Measures of motor activity revealed no group differences. Although all subjects gained weight on the free-food diet, the isolates did not gain any more weight than the ferals even though they had much higher caloric intake. The authors interpreted these data as evidence for a metabolic abnormality in the isolate subjects.

Another effect of social isolation during infancy, occurring in all mammalian species studied to date, is an abnormal reaction to painful stimulation (Melzack and Scott, 1957). Lichstein and Sackett (1971) assessed reaction to pain in 6- to 8-year-old rhesus monkeys that had been raised during infancy in social isolation. Compared with feral controls, isolates were much more reactive to very low, presumably non-painful levels of electric shock. However, isolates tolerated much higher levels of electric shock coupled with a metal drinking tube than did the ferals. And, surprisingly, when the drinking tube did not convey shock, isolates showed more generalized fearful behavior toward the tube than did ferals. These data suggest that the isolates had a major deficit in learning to develop adaptive reactions to painful stimulation, which may be based on a neurophysiological abnormality.

A study suggesting a major physiological abnormality in isolate rhesus monkeys was conducted by Sackett et al. (1973a). Two-year-old isolates and social controls were studied by radioimmune assay technique for cortisol response to stress. To measure maximum cortisol stress response, each subject was injected with ACTH and the 4-hr time course of cortisol change from basal level was measured. Subsequently, the subjects were all grouped together in a stressful social setting and groups were compared for basal cortisol versus post-playroom changes. This test was certainly stressful as none of the monkeys had previously been in a playroom, and none had been paired with more than one animal at any time. ACTH injection was found to stress both isolates and socialized controls equally, suggesting that the adrenal systems of both groups were equally capable of large cortisol rises. However, isolates had twice the basal level of cortisol (40 µ g%) observed in the controls. After the playroom stress test, the socialized monkeys showed the expected rise in cortisol, but the isolates failed to show any rise above basal value at the four 1-hr intervals after the playroom experience. Thus, the isolation rearing experience seems to have produced an abnormal adrenal stress response system, and experimenters investigating this system might be misled if they used such subjects to generate data.

CONCLUSIONS

It seems likely that after reading a chapter such as this, most colony managers might simply throw up their hands in despair. We have described a number of phenomena suggesting that both drastic and subtle but important aspects of behavior and physiology can be influenced by the way we rear and house our monkey populations, but we have not proposed solutions to most of the issues raised. Some remedies are easily pointed to: do not rear monkeys in social isolation; do not house potential social group members next to each other in single cages for long periods before forming social groups. Other problems are much more difficult and may be impossible to solve under the housing and colony-size conditions of a particular installation; they demand much more research data before a rational (i.e., non-trial-and-error) approach can be taken.

Our main hope is that the type of material presented in this chapter will produce an awareness and appreciation of the fact that differences between colony monkeys can be accounted for by more precise concepts than "individual differences," and, conversely, that we can and should account for some of these differences. Some of the causes for differences between animals are known, and others can be identified among variables existing in colony procedures and in colony records (e.g., Sackett, Holm, Davis, Fahrenbruch, 1975). We hope that the manager who is responsible for the health and breeding success of a colony will also be aware of the suitability of various types of monkeys for particular types of research projects. This could involve directing investigators to sources of information leading to appropriate choice of particular monkeys for subjects, rather than automatically putting "expendable" animals into invasive medical research.

REFERENCES

Alexander, B. K. The effects of early peer deprivation on juvenile behavior of rhesus monkeys. Unpublished doctoral dissertation, University of Wisconsin, 1966.

Arling, G. L. and Harlow, H. F. Effects of social deprivation on maternal behavior of rhesus monkeys. J. comp. physiol. Psychol. 64:371-377, 1967.

Arling, G. L., Ruppenthal, G. C., and Mitchell, G. D. Aggressive behaviour of the eight-year-old nulliparous isolate female monkey. Anim. Behav. 17:109-113, 1969.

Castell, R. and Wilson, C. Influence of spatial environment on development of mother-infant interaction in pigtail monkeys. Behaviour 39:202-211, 1971.

Griffin, G. A. The effects of multiple mothering on the infant-mother and infant-infant affectional systems. Unpublished doctoral dissertation, University of Wisconsin, 1966.

Hanson, E. W. The development of maternal and infant behavior in the rhesus monkey. Behaviour 27:107-149, 1966.

Harlow, H. F. The nature of love. Amer. Psychol. 13:673-685, 1958.

Harlow, H. F. and Harlow, M. K. The affectional systems. Pp. 178-185 in: Behavior of Nonhuman Primates, Vol. 1. Schrier, Harlow and Stollnitz (Eds.), New York: Academic Press, 1965.

Jensen, G. D., Bobbitt, R. A. and Gordon, B. N. Studies of mother-infant interactions in monkeys. Proc. 2nd Int. Congr. Primatol. 1:186-193, 1969.

Jensen, G. D., Bobbitt, R. A., and Gordon, B. N. Dominance testing of infant pigtailed monkeys reared in different laboratory environments. Proc. 3rd Int. Congr. Primatol. 3:92-99. 1971.

Lichstein, L. and Sackett, G. P. Reactions by differentially raised rhesus monkeys to noxious stimulation. Develop. Psychobiol. 4:339-352, 1971.

Melzack, R. and Scott, T. H. The effects of early experience on the response to pain. J. comp. phsysiol. Psychol. 50:155-161, 1957.

Miller, R. E., Caul, W. F., and Mirsky, I. A. Patterns of eating and drinking in socially-isolated rhesus monkeys. Physiol. Behav. 7:127-134, 1971.

Mitchell, G. D., Arling, G. L., and Møller, G. W. Long-term effects of maternal punishment on the behavior of monkeys. Psychonom. Sci. 8:209-210, 1967.

Mitchell, G. D., Harlow, H. F., Griffin, G. A., and Møller, G. W. Repeated maternal separation in the monkey. Psychonom. Sci. 8:197-198, 1967.

Mitchell, G. D., Ruppenthal, G. C., Raymond, E. J., and Harlow, H. F. Long-term effects of multiparous and primiparous monkey mother rearing. Child Develop. 37:781-791, 1966.

Pratt, C. L. and Sackett, G. P. Social partner selection in rhesus monkeys as a function of peer contact during rearing. Science 155:1133-1135, 1967.

Ruppenthal, G. C., Arling, G. L., Harlow, H. F., Sackett, G. P., and Suomi, S. J. A 10-year perspective of motherless-mother monkey behavior. J. abnorm. Psychol. 85:341-349, 1976.

Sackett, G. P. Some persistent effects of differential rearing conditions on preadult social behavior of monkeys. J. comp. physiol. Psychol. 64:363-365, 1967.

Sackett, G. P. Exploratory behavior of rhesus monkeys as a function of rearing experiences and sex. Develop. Psychol. 6:260-270, 1972.

Sackett, G. P., Bowman, R. A., Meyer, J. S., Tripp, R., and Grady, S. Adrenocortical and behavioral reactions by differentially raised rhesus monkeys. Physiol. Psychol. 1:209-212, 1973.

Sackett, G. P., Griffin, G. A., Pratt, C., Joslyn, W. D., and Ruppenthal, G. C. Mother-infant and adult female choice behavior in rhesus monkeys after various rearing experiences. J. comp. physiol. Psychol. 63:376-381, 1967.

Sackett, G. P., Holm R. A., Davis, A. E., and Fahrenbruch, C. E. Prematurity and low birth weight in pigtail macaques: Incidence, prediction, and effects on infant development. Pp. 189-206 in: Proceedings of the Fifth Congress of the International Primatological Society. Tokyo: Japan Society Press, 1975.

Sackett, G. P. and Ruppenthal, G. C. Development of monkeys after varied experiences during infancy. Pp. 52-87 in: Ethology and Development. S. A. Barnett (Ed.), London: Spastics International Medical Publication, 1973a.

Sackett, G. P. and Ruppenthal, G. C. Induction of social behavior changes in macques by monkeys, machines, and maturity. Pp. 99-120 in: Fourth Western Symposium on Learning: Social Learning. P. J. Elich (Ed.), Bellingham, WA: Western Washington State College, 1973b.

Seay, B. Maternal behavior in primiparous and multiparous rhesus monkeys. Folia primat. 4:146-168, 1966.

Singh, S. Urban monkeys. Sci. Amer. 221:108-115, 1969.

Suomi, S. J. and Harlow, H. F. Social rehabilitation of isolate-reared monkeys. Develop. Psychol. 6:487-496, 1972.

V. CARE OF EXOTIC SPECIES

CHAPTER 21

HAND-REARING INFANT GIBBONS

A. W. Breznock,* S. Porter,[+] J. B. Harrold,* and
T. G. Kawakami*

*Comparative Oncology Laboratory
University of California
Davis, California

[+]Gladys Porter Zoo
Brownsville, Texas

INTRODUCTION

Nursery rearing of nonhuman primates has become a standard and accepted procedure for obtaining neonatal animals for experimentation and for sustaining the life of a rejected infant (Jacobson and Windle, 1960; Blomquist and Harlow, 1961; Fleischman, 1963; Vice, Fritton, Ratner, and Kalter, 1966; Hall, 1975). The techniques first described by Fleischman for hand-rearing macaques have been used as a guideline and modified to suit the needs of other species, such as baboon, marmoset, lemur, bushbaby, and chimpanzee (Okano, 1962; Vice et al., 1966; Hampton and Hampton, 1967; Hall, 1975). These initial techniques have also been used as the basis for hand-rearing infant gibbons (Hylobates lar) with the additional considerations of longer dependency period, contact with humans, and exercise requirements (Carpenter, 1940; Sasaki, 1962; Berkson, 1966; Ellefson, 1967; Arnold, 1973).

The only reason for hand-rearing gibbon infants, thus far, has been a threat to the infant's life, either from maternal rejection or disease. This chapter contains observations on the hand-rearing of four maternally rejected infant gibbons in laboratory and zoo environments. Both the laboratory and the zoo were striving for normal offspring that could sustain a breeding colony and also be used for display. The ultimate goal was to simulate natural rearing and to rear the infants with the mother to at least 5 months of age in the laboratory, or for as long as the primary family would accept the infant, usually 3-4 years in the zoo (Carpenter, 1940; Harlow and Harlow, 1962; Ellefson, 1967; Berkson, Ross, and Jatinandana, 1971; Arnold, 1973). The rate of weight gain, disease conditions, and the behavior and activity patterns of colony-born, mother-reared offspring were used as criteria to assess the development of the hand-reared gibbon (van Wagenen and Catchpole, 1956; Mason, 1965). The data show that the hand-reared babies appear normal in their physical development but display behavioral abnormalities that may be due to deficiencies in the hand-rearing procedures.

MATERIALS AND METHODS

Infant 1 was one of nine infants born at the Comparative Oncology Laboratory at the University of California, Davis. It entered the nursery at 27 days of age after being severely injured by its mother. Infants 2 and 3 (H. l. pileatus) were born at the Gladys Porter Zoo in Brownsville, Texas, and entered the nursery on days 6 and 1 postpartum. Infant 4 entered the nursery of the Oregon Wildlife Safari on day 6 after birth. Gestation ranged from 190 ±7 to 214 ±7 days, and all infants were delivered vaginally.

All infants were housed in standard isolette incubators (Armstrong Baby Incubator, Model 500, Gordon Armstrong Co., Cleveland, OH) with towels or shag rug rolled up for clinging. The temperature was maintained at 88-95° F during the first 6 weeks of age, and humidity was kept at 65%.

All infants readily accepted formula (Probana and Enfamil, Mead Johnson and Co., Evansville, IN) and premature infant or Pet-Nip nurser nipples (Poly-Nurser Products Corp., Brooklyn, NY). Formula intake began at 7-10 ml per feeding and was increased to a maximum of 35 ml per feeding at 5 months of age. A vitamin supplement (Vi-Daylin, Ross Laboratories, Columbus, OH), 2 drops per day, was added to the formula. Infant 1 preferred feeding from a 12-cc plastic syringe with no nipple after 14 days. All infants were taken out of the incubator and fed sitting upright with towels wrapped around their bodies. They were encouraged to stand, grasp, and exercise their legs at each feeding. They were burped by rubbing the back after feeding. Prepared baby foods were fed in addition to the formula at 2-2½ months to all infants. Fruits, cereals, and softened monkey chow were offered at 3-3½ months of age.

Infant 1 was fed 5-6 times a day for the first 3 months in the nursery, with no feedings between 11 p.m. and 7 a.m. Infants 2 and

3 were fed 8 times evenly spaced in 24 hr. Infant 4 was fed on demand, averaging at 2-hr intervals around the clock until 11 weeks of age, when it began sleeping for 7-8 hr through the night.

Weights were recorded daily until 3 months of age for infant 1, at least twice a week for infant 3, and liquid food intake was recorded for infants 1, 2, and 3 until weaning at 6 months of age. Rectal cultures and stool examinations for intestinal parasites were done upon entry into the nursery, and at any time that diarrhea occurred.

At 3 months, infant 1 was put in a larger cage equipped with a series of plastic bars for a few hours each day. At 4 months he was housed permanently in the large cage and at 6 months a mother-reared infant was housed with him. Infants 2 and 3 were removed from the nursery at 5 and 4 months respectively to a large display pen enriched with branches. Infant 2 was housed first with a 2-year-old male gibbon and then with a 4-month-old female mandrill. Infant 4 received almost continual human contact from 6 weeks of age. When it was not being carried by a person, it clung vertically to a roll of shag rug which was rocked frequently in the incubator. At 3 months it spent several hours a day in a larger pen with a bar apparatus, but it was returned to the incubator at night. This routine is still employed at 5 months of age, and no cagemate has become available. No behavioral experimental design was devised to assess infant development but observations of their reactions in relation to mother-reared peers were recorded.

Mother-reared weanlings (5-8 months) at the Comparative Oncology Laboratory are housed in pairs or triplets when possible (Harlow and Harlow, 1962; Badham, 1967). They are placed in a double Norwich-type cage equipped with perches and swing bars. They are fed two bottles of milk per day, one bottle of Gatorade (Stokely Van Camp Inc., Indianapolis, IN), soaked monkey chow (Wayne's Monkey Diet, Allied Mills Inc., Chicago, IL), fruit, and baby foods. Food intake is recorded daily and the animals are weighed at least once a month. Weanlings at the zoo and wildlife safari are housed in the larger cages previously described.

RESULTS

Initial weights of gibbon infants upon entry into the nursery were 490 g, 336 g, 462 g, and 356 g, respectively for infants 1, 2, 3, and 4. Weight gains ranged from 1.9 to 9.8 g/day for the four infants during the 5 months of nursery rearing, with a mean of 5.3 g/day and a standard deviation of ±0.98 (Figure 1).

Highest consistent rates of gain were obtained by infants 2 and 3, which were fed eight times per day. Infant 4 gained an average of 9.8 g/day for the first month, while being fed on demand; rate of gain declined thereafter to 5.6 g/day for the second month, 4.1 g/day for the third month, 2.5 g/day for the fourth month, and 1.9 g/day for the fifth month. Infant 1 gained an average of 4.1 g/day for the 5-month period on 5–6 feedings per day. The hand-reared infants all weighed at least 1 kg at weaning from the nursery (5 months of age), compared with an average weight of 900 g at 5 months for four mother-reared gibbons. Rates of gain from weaning to 1 year of age averaged 3.8 ± 1.28 S.D. per day for four mother-reared weanlings and 3.44 ± 1.8 S.D. per day for one hand-reared weanling (Figure 2).

Hand-reared and mother-reared infants showed active interest in solid foods at about 3 and 4 months respectively (Table 1), and both displayed a marked preference for fruit-flavored foods over commercial monkey chow or protein cereals. Self-feeding from a hanging bottle was encouraged at 4½ months of age in the hand-reared infants and was achieved after 7 days of training in which the infant was held up to the feeder at the regular feeding times. Mother-reared weanlings readily adapted to the hanging bottles within 24 hr of weaning.

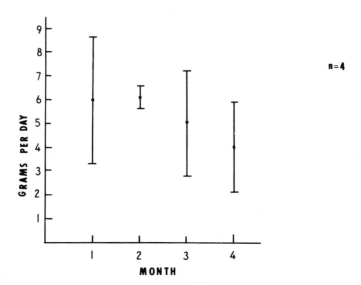

Figure 1.   Rate of weight gain in hand-reared gibbons.

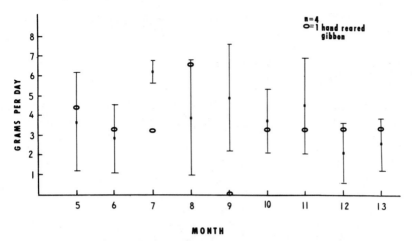

Figure 2.   Rate of weight gain in mother-reared weanling gibbons.

TABLE 1.   DEVELOPMENT OF MOTHER-REARED AND HAND-
REARED GIBBON INFANTS

ACTIVITY	MOTHER-REARED (n=7)	HAND-REARED (n=variable)
Crawl with head down	never	3 wks (n=3)
Lifted head and supported body by elbows	day 1	6- 8 wks (n=4)
Turned over	?	8 wks (n=3)
Kneeling, sitting, or squatting	6-8 wks	8-10 wks (n=3)
Stood up alone	8-10 wks	10-12 wks (n=4)
Climbed vertically	8-10 wks	12-14 wks (n=4)
Swinging by both hands	14-16 wks	18-10 wks (n=4)
Ate solid food	16-20 wks	10-15 wks (n=4)
Sucking thumb	never	2- 8 wks (n=4)
Self clasping and rocking	never	4 wks (n=2)
Ventral recumbancy response	never	12 wks (n=2)

Infant 1 was severely injured by its mother and was in shock when found on the floor of the cage. Its body temperature was 94° F. Warm intravenous lactated Ringers solution was given at a slow rate for a total of 35 ml with 50 mg of prednisolone sodium succinate (Solua-Delta Cortef, Upjohn Co., Kalamazoo, MI) added. Two hours later 10 ml of dextrose was given orally which was continued every 2 hr for 24 hr. Staphylococcus aureus was cultured from an abscess that invaded the elbow joint. Body temperature remained at or below 101° F, but the axillary lymph nodes became markedly enlarged, to 4 mm in diameter. A complete blood count 24 hr after extension of the abscess into the elbow joint showed a WBC of 7,500, with a differential of 21% band neutrophiles, 25% segmented neutrophiles, 48% lymphocytes, 5% monocytes, 0% basophiles, and 1% eosinophiles, 7.0 million RBC, 14.3 g% hemoglobin, 42% hematocrit, and 6.0 g% plasma proteins. Normal CBC values for weanlings between 6 and 12 months of age have an increased percent of lymphocytes compared with neutrophiles (Table 2). Oral Ampicillin was given 4 times daily for a total dose of 50 mg/kg/day. After 3 days of treatment, diarrhea and dehydration were noted. The antibiotic was changed to Erythromycin oral suspension at 40 mg/kg/day divided 3 times daily for 7 days. Solid stool resulted 24 hr after the antibiotic was changed. All abscesses were treated locally with penicillin ointment, hydrogen peroxide, and warm soaks for 15 min 3 times a day. The abscesses took 50 days to heal completely. The infant's food intake averaged 125 ml/day during the treatment and weight continued to increase, except when diarrhea occurred (Figure 3).

Enteritis, with Shigella flexneri type 6 isolated, occurred in a mother-reared infant one week after weaning. Anorexia and depression preceded the diarrhea by 12 hr. The infant had profuse liquid stool for 24 hr and semisolid unformed stool for another 48 hr. Its temperature was normal at 100.4° F. When anorexia and depression were noted, oral treatment with chloramphenicol was initiated at a dose of 60 mg/kg divided 4 times per day. Hand feedings of 25 ml of Sustagen (Mead Johnson and Co., Evansville, IN) and an oral potassium preparation (Kaon Elixir, Warren Teed, Columbus, OH) (2.0 meq/kg/day) 4 times a day were initiated when diarrhea was first noted, and 50 ml of Ringers lactate was given subcutaneously twice a day. Treatment lasted for 48 hr, at which time appetite returned and stool became semisolid. Chloromycetin treatment was continued for 7 days. A stool culture 10 days later was negative for Shigella.

TABLE 2. HEMATOLOGICAL VALUES FOR THREE AGE GROUPS OF GIBBONS

Compiled January, 1977, by T. Kawakami and B. Harrold

	N	WBC	BAND	SEG.	LYMPH	BASO	MONO	EOS	RBC	HCB	HCT	PP
Normal Adults 4 yr	502	9,991 3,792	104 214 0.92%	5,236 2,996 50%	3,947 1,605 41.6%	336 376 3.4%	99 139 1.05%	216 238 2.2%	7.2 0.8	15.1 2.17	45.7 5.6	6.4 0.5
Normal Juveniles 1–4 yr	13	8,875 3,468	151 170 1.5%	2,746 2,227 29.9%	5 487 2,249 63%	229.6 154 2.7%	177 265 1.8%	83 103 0.8%	6.5 0.5	13.88 1.3	38 3.5	6.0 0.4
Normal Weanlings 0.5–1 yr	20	9,620 3,440	146 168 1.5%	2,955 1,765 30.7%	5,890 2,164 61.2%	242 123 2.5%	140 168 1.4%	124 123 1.2%	6.9 0.83	13.5 1.28	39 4.0	6.1 0.5

A second weanling displayed anorexia and a liquid stool that was positive for occult blood at 3 weeks after weaning. A stool sample showed infection with <u>Giardia</u>. Flagyl (Searle and Co., San Juan, PR) was administered by stomach tube twice a day for 3 days at a dose of 30 mg/kg/day. The stool became firm after 48 hr of treatment, and stool samples 3 days and 2 weeks after treatment

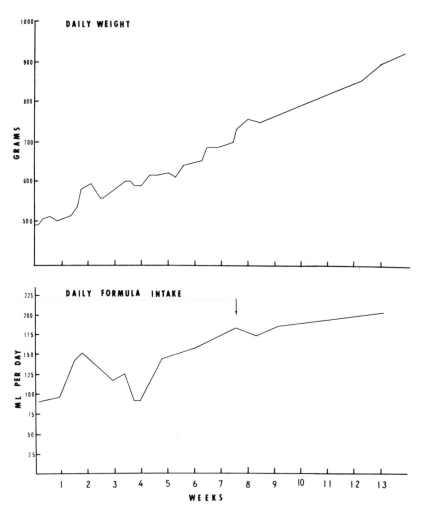

Figure 3.   Daily weight and formula intake for hand-reared gibbon #1.

were negative for Giardia. There has been no recurrence of the parasite nor has the cagemate been infected.

Infant 4 experienced severe diarrhea at 3 weeks and was anorexic for 2 days. A stool culture showed a non-lactose-fermenting E. coli. Chloromycetin palmitate was given for 7 days and the stool became normal after 10 days.

Stereotypic behaviors have been noted in all four hand-reared infants, including thumb or wrist sucking, self-clasping, rocking, head banging and shaking, and ventral recumbancy positioning (Table 1). Infants 1 and 2 repeatedly position themselves on the abdomen and clasp the perineal region with one hand while sucking the thumb or wrist of the other hand. They respond in this manner to the sight of unfamiliar humans, or to sudden movements of their cagemates. Infant 1 shakes and bangs his head. While these infants sit unstimulated, they suck their thumbs or wrists. Infant 1 began perineal clasping immediately upon entering the nursery at 27 days of age; he began thumb sucking at 2 months of age, but the head shaking was not observed until he was weaned and confronted with a peer. As soon as he could stand, at 3 months, it was noted that when an unfamiliar person entered the room he fell into a prone position and clasped his perineal region. When this infant was weaned and placed with a mother-reared peer, he avoided physical contact vigorously, shrieking and falling to the abdominal position when touched. Mother-reared infants, when first paired with peers, do not hesitate to touch, explore, vocalize softly and clasp each other in manners that have been previously described (Bernstein and Schusterman, 1964; Paluck, Lieff, and Esser, 1970). Infant 2 reacted identically to infant 1 when it was placed with a 2-year-old gibbon at weaning and later with a mandrill. Infant 3 displayed wrist sucking, rocking in a sitting position, and self-clasping. Infant 4 displayed thumb sucking at 2 weeks of age but ceased by 14 weeks. This infant is now 5 months old and displays no stereotypies. In all hand-reared infants, the development of motor abilities such as standing up, climbing, and swinging was behind that of mother-reared infants by at least 2-4 weeks.

DISCUSSION

As seen in other nonhuman primates, hand-rearing of infant gibbons results in good weight gains, equal to or above those of mother-reared infants (Fleischman, 1963; Berkson, 1966; Carpenter, 1940; van Wagenen and Catchpole, 1956). The birth weight of gibbons (440-505 g for 8 infants) is comparable to that of rhesus monkeys (465 ±70 g S.D.), and the adult weights are also comparable, so the rhesus rates of gain, 3.3-4.0 g per day, were held to be a reasonable standard (van Wagenen and Catchpole, 1956; Jacobson and Windle, 1960; Fleischman, 1963; Hall, 1975). Therefore, the techniques described here for hand-rearing infant gibbons provide good weight gains and healthy animals. Medical problems can be dealt with through careful observation and prompt action to insure normal hydration and electrolyte balance.

The major difficulty seems to be that the hand-reared infants at the Comparative Oncology Laboratory and the Gladys Porter Zoo did not have a proper social environment to develop into normal individuals. The infant gibbon is dependent on its mother for a comparatively long period, about 1-1½ years (Badham, 1967; Arnold, 1973; Brody and Brody, 1974), and derives stimulation, discipline, and support from this long and close attachment (Berkson, 1966; Arnold, 1973). Our laboratory-rearing environment does not provide this stable and stimulatory effect.

The stereotypies observed in three of these hand-reared gibbons fall into the same types reported in socially-deprived isolate-reared macaques (Harlow and Harlow, 1962; Mason, 1965). Behavior such as thumb sucking has been rarely observed in wild-born macaques and would be even more doubtful to occur in a species whose mode of locomotion is brachiation at a height sometimes exceeding 80 ft (Mason, 1965; Carpenter, 1940; Ellefson, 1967). This type of behavior has not been described in field reports on natural gibbon activity (Berkson, 1966; Ellefson, 1967). An infant must spend much of its time clinging vertically by both hands and feet to its mother as she brachiates or forages. The mother rarely supports the infant with her arms (Carpenter, 1940; Mason, 1965). The abnormal pattern did not occur in 7 colony-born mother-reared infant gibbons.

With the occurrence of extreme behavior abnormalities in infant 1, the frequency and amount of physical and social activity were examined and thought to be deficient, although adequate for growth and health. The animal was handled only 6 times a day and left alone to cling to a towel the remainder of the time. The infant usually ended up in a horizontal position, off the towel and clinging only to himself. This environment lacked the stimulation of movement and animal contact. Infant 4, on the other hand, was provided with a vertical, moving orientation as well as more physical and social contact, and this infant displayed only transient thumb sucking, with no other stereotypies to date. Since the 24-hr personal care method used with infant 4 is impractical in a laboratory situation, a swinging surrogate covered with shag rug and mounted with a bottle should be considered if socially normal infants are desired.

Gradual socialization with peers is ideal for most hand-reared species (Harlow and Harlow, 1962; Mason, 1965). Small gibbon breeding colonies may not allow socialization with a peer, but early and gradual socialization beginning at 1-2 months of age with a nonhuman primate of any species would appear to be more desirable than no contact for the first 5-6 months of age. Infant 2 has been able to coexist well with a mandrill baboon, although at 2 years of age the permanent effects of its abnormal behavior on future adult interactions with other gibbons cannot be determined. Social contact with humans may not be detrimental if supplemented with gibbon contact later. Young "pet" gibbons have entered the Comparative Oncology Laboratory colony, reached maturity, and proven to be viable sires and dams, although the Gladys Porter Zoo has had the opposite experience with pet gibbons. There may be a critical age at which co-species exposure is necessary, and after which time rehabilitation is not possible. Infant 1 continues to exhibit abnormal patterns even though he has been housed with 2 mother-reared peers for 10 months.

ACKNOWLEDGMENTS

We are indebted to the nursery-care staff at Gladys Porter Zoo, the Oregon Wildlife Safari, and the Comparative Oncology Laboratory for their expert and faithful assistance in raising the infants. In addition to her animal care responsibilities, the excellent recording of data by Lori Marker at the Oregon Wildlife Safari is gratefully appreciated. This work was supported by grant #NO1-CP-3-3242 from the Comparative Oncology Laboratory, University of California at Davis.

REFERENCES

Arnold, R. C. Births of gibbons in captivity. Gibbon and Siamang 2:221-227, 1973.
Badham, M. A. A note on breeding the pileated gibbon. Int. Z. Yearbook 7:92-93, 1967.
Berkson, G. Development of an infant in a captive gibbon group. J. genet. Psychol. 108:311-325, 1966.
Berkson, G., Ross, B., and Jatinandana, S. The social behavior of gibbons in relation to a conservation program. Primate Behavior 2:226-255, 1971.
Bernstein, I. S. and Schusterman, R. J. The activity of gibbons in a social group. Folia primat. 2:161-170, 1964.

Blomquist, A. J. and Harlow, H. F.   The infant rhesus monkey program at the University of Wisconsin primate laboratory. Proc. anim. Care Panel 11:57-64, 1961.

Brody, E. J. and Brody, A. E.   Breeding Muller's Bornean gibbon. Int. Z. Yearbook 14:110-113, 1974.

Carpenter, C. R.   A field study in Siam of the behavior and social relations of the gibbon. Comp. Psychol. Monogr. 16(5), 1940.

Ellefson, J. O.   A natural history of gibbons in the Malay Peninsula. Ph.D. Dissertation, University of California, Berkeley, 1967.

Fleischman, R. W.   The care of infant rhesus monkeys (Macaca mulatta). Lab. Anim. Care 13:703-709, 1963.

Hall, A.   Nursery procedures at the Oregon Regional Primate Research Center. Prim. News 13:2-6, 1975.

Hampton, S. H. and Hampton, J. K.   Rearing marmosets from birth by artificial laboratory techniques. Lab. Anim. Care 17:1-10, 1967.

Harlow, H. F. and Harlow, M. K.   Social deprivation in monkeys. Sci. Amer. 207:137-146, 1962.

Jacobson, H. N. and Windle, W. F.   Observations on mating, gestation, birth, and postnatal development of Macaca mulatta. Biol. Neonat. 2:105-120, 1960.

Mason, W. A.   The social development of monkeys and apes. Pp. 514-543 in: Primate Behavior, I. DeVore (Ed.), New York: Holt, Rinehard, and Winston, 1965.

Okano, T.   Experimental raising of an infant chimpanzee. Primates 3:75, 1962.

Paluck, R. J., Lieff, J. D., and Esser, A. H.   Formation and development of a group of juvenile Hylobates lar. Primates 11:185-194, 1970.

Sasaki, T.   Handrearing a baby gibbon (Hylobates lar). Int. Z. Yearbook 4:289-290, 1962.

van Wagenen, G. and Catchpole, H. R.   Physical growth of the rhesus monkey (Macaca mulatta). Amer. J. phys. Anthrop. 14:245-273, 1956.

Vice, T. E., Britton, H. A., Ratner, I. A., and Kalter, S. S.   Care and raising of newborn baboons. Lab. Anim. Care 16:12-22, 1966.

CHAPTER 22

CARE OF THE INFANT AND JUVENILE
GIBBON (Hylobates lar)

D. P. Martin, P. L. Golway, M. J. George, and
J. A. Smith

Department of Laboratory Animal
 Medicine and Science
Litton Bionetics, Inc.
Kensington, Maryland

BACKGROUND

A breeding colony of six female and seven male gibbons
(Hylobates lar) has been maintained at Litton Bionetics, Inc. (LBI)
since July 1973, when they were obtained as a group from a
pharmaceutical firm. At that time, all animals in the colony were
mature or nearly so and had been in the United States for 2-5
years. The first attempts at breeding these animals began 10
months after they arrived at our facility, and in the intervening 2
years, 12 pregnancies have been diagnosed among the six females
currently in the colony. These 12 pregnancies have resulted in six
live births (two males and four females) with two cases of fetal
wastage and four animals still pregnant.

This colony is maintained for the purpose of infant production. Our goal is to produce healthy animals that can be used for
research and/or as breeders when they reach adulthood.

PERINATAL CARE

All infants in the colony were delivered vaginally and, at this
date, range in age from less than 2 months to 29 months. Three of
the infants, including both the youngest and the oldest, have been
entirely mother-reared. One animal, delivered prematurely, was
entirely hand-reared, and two animals were mother-reared for 8-11

weeks, then nursery-reared until weaning from the bottle. The decision to separate these latter two animals from their mothers was made in one case following a severe injury to the infant's hand and in the other case when an infant failed to gain weight properly and an examination of the mother revealed scant, watery lactation.

Between 2 and 9 days postpartum, each mother was restrained chemically with phencyclidine hydrochloride (Serny-lan(R), Bio-Ceutic Laboratory, Inc., St. Joseph, MO) or ketamine hydrochloride (Ketaset(R), Bristol Laboratories, Syracuse, NY). At this time, the infant was examined and weighed, and the mother examined for the presence of retained placenta, the state of uterine contraction, and any other evidence of abnormality.

The mean birth weight or, more correctly, "first weight" for the infants was 370 g with a range of 264-437 g. Based on records of mating times, the gestation periods for these infants agree well with the literature, approximately 210 days. The mating records for the parents of the smallest infant indicated that this animal was born 1 month prematurely; the scant hair coat and the weight of 264 g confirmed this fact. If this weight is eliminated from the calculations, the remaining five animals have a mean birth weight of 391 g.

DIET

Once the animals were brought to the nursery, they were treated in much the same manner as we have reported for macaque rearing. A modified milk substitute was made from a combination of 520 g of a commercially prepared human infant formula (SMA(R) Powder, Wyeth Laboratories, Philadelphia, PA) and 210 g of lactose mixed in 3 liters of water. One to two drops of a vitamin supplement (Paladac(R), Parke Davis, Detroit, MI) were added to the first bottle of the day. The LBI nursery is staffed between 7:00 a.m. and midnight, although a night technician is available until 3:00 a.m. for late night feedings. In contrast to macaques, which can usually be taught to self-feed from a fixed bottle in 3-7 days, the gibbon that was nursery-reared from birth did not start self-feeding until 105 days. One partially mother-reared animal brought to the nursery at 76 days was self-feeding by the next day, and the other one, brought to the nursery at 60 days, did not self-feed regularly until 122 days. This latter animal may have been able to self-feed earlier than this; the records show that the animal did self-feed for 2-3 days beginning at day 62.

The amount of formula consumed and the feeding schedule varied somewhat among animals, but feedings of 2-3 ml were offered every 2-3 hr, and the volume was gradually increased while the number of feedings was decreased. At 80 days of age, all three animals housed in the nursery were regularly consuming 100-140 cc of formula each day. Baby food was mixed with the formula to thicken it. At approximately 4½ months of age, the nursery-reared animals began consuming a mush made from three parts of biscuits (Monkey Chow(R), Ralston Purina Company, St. Louis, MO) soaked in one part of water and one part of applesauce. Perhaps not so coincidently, we observed that mother-reared infants began to eat fruit at about 4 months of age.

WEIGHT GAIN

Weight gains among the infant gibbons during the first 6 months of life are illustrated in Figure 1. This figure illustrates that whether the animal is nursery-reared, mother-reared, or a combination of both, weight gains are fairly constant. The failure to gain weight properly while nursing on the mother can be clearly seen in infant B-7192. Also illustrated are episodes of clinical disease in two infants while in the nursery (B-7669 and B-6997) which affected appetite and, consequently, weight gains.

In Figure 2 the time frame is expanded to 1 year and the figure depicts only those three infants that have been alive for that period. Again, weight gains are fairly constant. Also illustrated is the increase in rate of weight gain which is consistent with the time that the infants began to eat solid food. The mean daily weight gain for the infants for the first 6 months of life was 2.38 g. For the second 6 months the mean gain was 3.84 g per day, and the mean gain for the entire year was 3.12 g per day.

HOUSING

Infant gibbons brought to the nursery are first housed in a human infant incubator in which temperature, humidity, and oxygen flow can be adjusted. Once they are stable, they are moved to wire cages (60 x 48 x 40 cm) with diapers so that movement and brachiation are encouraged. At the age of 2-3 months, infants raised in the nursery are still moving by crawling about, and the bottle holder for self-feeding is located low in the cage to assure that the animal can reach it.

The next step for the hand-reared gibbon is to remove them to the adult gibbon housing area. The entire transfer cage is placed on the floor of one of the adult cages (1.0 x 1.2 x 1.5 m) and

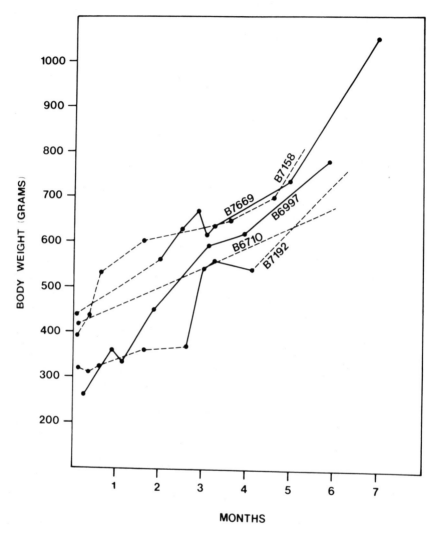

Figure 1.   Body weight gains of infant gibbons during the first 6
months of life.
mother-reared = --------   nursery-reared = ——————

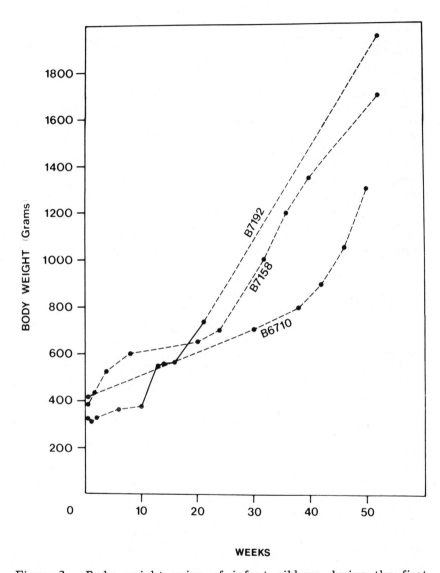

WEEKS

Figure 2.   Body weight gains of infant gibbons during the first
            year of life.
            mother-reared = --------   nursery-reared = ——————

the infant is allowed to adjust to the new sights and sounds of this colony. Eventually, the diapers are removed and the animal is given the use of the entire adult cage.

SOCIALIZATION

As might be expected from work in other species, nursery-reared gibbons are socially backward, and in order to increase the possibility of their becoming normal adults, we have encouraged socialization by housing nursery-reared animals with mother-reared peers whenever possible. These attempts may be unsuccessful if the mother-reared animal totally dominates the other. If a less dominant peer is selected and the amount of time that the two animals spend together is gradually increased, even if for only a few minutes each day at first, socially handicapped infants do seem to become more normal in their activities. Unsocialized gibbons, like other nursery-reared nonhuman primates, engage in stereotyped behaviors such as thumb sucking, head banging, rocking, etc.

WEANING

We have observed that to a certain extent the ease and speed with which a mother-reared infant is weaned depends upon the mother. For example, we have noted that the infant of a "good" mother that offers the infant small pieces of fruit, such as a piece of banana or grape, and actually helps the infant climb up the bars of the cage, progresses faster than the infant of a mother that keeps all the treats for herself and lets the offspring explore on its own. Although mother-reared infants have been weaned at about one year of age, observations with the nursery-reared animals indicate that weaning could occur much earlier. Early weaning may be desirable if maximum production is to be achieved. The female whose infant was separated at 76 days of age was immediately returned to breeding and gave birth to another full-term infant 218 days later.

CLINICAL CARE

All clinical problems in the infant and juvenile gibbons have been related to the enteric system. Like the adults of this species, these animals may react to small psychic stresses with diarrhea. Occasionally, a problem with diarrhea has become extreme enough to warrant anal culture and treatment. Pseudomonas aeruginosa,

Shigella flexneri, and Klebsiella spp have been cultured from these animals as well as several other less pathogenic bacteria. The bacteria cultured have been sensitive to a wide variety of antibiotics in vitro. They have responded well in vivo to colistin sulfate (Coly-Mycin S Oral Suspension(R), Warner-Chilcott Laboratories, Morris Plans, NJ) and to gentomycin sulfate (Gentocin(R), Schering Corporation, Kenilworth, NJ). The diarrhea has usually been accompanied by lethargy and inappetence and, as might be expected, resultant weight loss.

Care of the six infants reported here has enabled us to gain experience and confidence in the husbandry of the young of this species. We intend now to expand our pool of knowledge by gaining further information such as body measurements and normal hematology data on the infants of the four females currently pregnant as well as with these six.

ACKNOWLEDGMENTS

This research was supported by NIH contract #NO1 CP 6-1006 from the National Cancer Institute. This work was conducted according to the principles of animal care promulgated by the NAS-NRC (see "Guide for Laboratory Animal Facilities and Care," DHEW Publication No. (NIH) 73-23).

CHAPTER 23

HAND-REARING INFANT CALLITRICHIDS (Saguinus spp AND Callithrix jacchus), OWL MONKEYS (Aotus trivirgatus), AND CAPUCHINS (Cebus albifrons)

J. L. Cicmanec, D. M. Hernandez,
S. R. Jenkins, A. K. Campbell, and
J. A. Smith

Department of Laboratory Animal
  Medicine and Science
Litton Bionetics, Inc.
Kensington, Maryland

A nursery for nonhuman primates has been maintained at Litton Bionetics, Inc., for 14 years (Valerio, Darrow, and Martin, 1970), but care of New World primate infants in these facilities has begun only recently. Breeding colonies of selected New World primate species were established within the last five years for studies in viral oncology. Among the species represented, the largest numbers of infants produced are callitrichids (Saguinus spp and Callithrix jacchus), owl monkeys (Aotus trivirgatus), and capuchins (Cebus albifrons).

With the occurrence of serious disease and parental neglect, we found it necessary to hand-rear these small infants. Methods for hand-rearing callitrichids have been well defined (Hampton and Hampton, 1967; Wolfe, Ogden, and Deinhardt, 1972), and we adopted or modified these procedures for our callitrichid and owl monkey infants.

MANAGEMENT PRACTICES

Many of the unique problems associated with the care of these infants are related to their small size. Most callitrichids do not exceed 40 g at birth, and some C. jacchus infants weigh less than 30 g when born. Newborn owl monkeys weigh between 70 and 120 g, and capuchin infants range between 200 and 260 g at birth. The thermoregulatory mechanism of New World primates is not well developed at birth, so we house all hand-reared infants in incubators stabilized at $88-90^{\circ}$ F and 50% $\pm$ 5% relative humidity.

Their relatively high caloric requirement and limited gastric capacity present definite restrictions on feeding practices. We use a formula composed of 15 g Sustagen(R) and 15 g SMA(R) per 60 ml of sterilized water. This formula provides 1.6 calories and 44 mg protein per ml. The quantity of formula fed for each of the species at various ages is shown in Table 1. Callitrichids are fed with a Unipette(R) dilution apparatus, and owl monkeys and capuchins are fed with a doll feeding bottle.

A specific nutritional requirement of New World primates is vitamin D_3. The SMA provides 12 units of vitamin D_3 per ml, but we also supplement the diet of infants with 50 I.U. per week.

Several problems can develop during feeding. Occasionally, infants eat too rapidly and regurgitate their food shortly after feeding. Once regurgitated, the ingesta can be aspirated into the lungs. If an infant frequently regurgitates, we thicken the formula with a small quantity of semi-solid baby food, which often reduces the frequency of regurgitation.

TABLE 1. INFANT FEEDING SCHEDULES

	Callitrichids	
Birth	0.3 - 0.5 cc	Every 2 hr
Day 10	0.7 - 1.0 cc	"
Day 25	1.5 - 2.0 cc	"
	Owl monkeys	
Birth	0.9 - 1.5 cc	Every 2 hr
Day 10	1.5 - 2.5 cc	"
Day 25	3.0 - 4.5 cc	"
	Cebus monkeys	
Birth	3.0 - 6.0 cc	Every 3 hr
Day 10	10.0 -15.0 cc	"
Day 25	20.0 -35.0 cc	"

We weigh infants frequently, as weight is an important determinant of health status. Typical weight gains for owl monkey and cebus monkey infants are shown in Figures 1 and 2.

CLINICAL DISEASE

The most common clinical diseases we encounter in infant New World primates are candidiasis and pneumonia. We also see traumatic injuries caused by the parents, including bitten tails, hands, and feet. Dermatitis and diarrhea also occur frequently.

Classical descriptions of candidiasis have been reported in Old World primate infants, and the appearance of the disease is similar in New World primate infants. The initial sign is a moist,

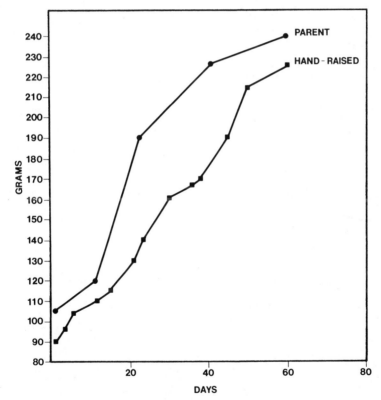

Figure 1.   Comparison of weight gain between a parent-reared and a hand-reared owl monkey infant.

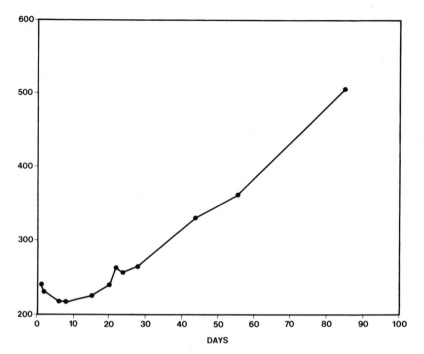

Figure 2.   Weight gain of a Caesarean-delivered, hand-reared
cebus monkey infant.

glistening glossitis which may progress and involves 70-90% of the
dorsal surface of the tongue.   In very rapidly spreading infections,
buccal surfaces of the mouth, lips, larynx, and esophagus may be
involved.   Anorexia and dehydration occur if cases are not rapidly
and effectively treated.   Parenteral treatment with mystatin
(Mycostatin oral suspension, E. R. Squibb and Sons, Inc., Princeton,
NJ) is 95% effective but must be continued for at least 10 days; in
many difficult cases treatment extends beyond 25 days.

    Most pneumonias are of bacterial origin, and the pathogens
most frequently isolated are Streptococcus pneumoniae, Klebsiella
pneumoniae, alpha-hemolytic Streptococcus, Staphylococcus aureus, and Pseudomonas aeruginosa. Early detection is paramount in
these species because infections rapidly progress and can easily
become fatal.   We find palpation of the chest a more useful
indicator of pulmonary malfunction than auscultation.   The small
size of the infants and the stressful nature of performing radiographs or blood collections preclude the use of these diagnostic

aids. We have found rate and rhythm of respiration to be so variable that these parameters also are not useful in diagnosing pneumonia of New World primate infants. Treatment with appropriate antibiotics, determined by in vitro culture of throat swabs in addition to fluids and bronchodilators, is indicated. Supplemental oxygen also aids in recovery. Some antibiotics that we have found particularly useful are ampicillin (Omnipen, Wyeth Laboratories, Philadelphia, PA), erythromycin (Ilosone, Dista Products, Indianapolis, IN), and chloramphenicol (Chloromycetin palmitate, Parke-Davis, Detroit, MI), all of which can be given orally. In some resistant infections, gentamycin (Gentocin, Schering Corporation, Chicago, IL) is given intramuscularly.

Whenever possible, medications are given orally rather than by intramuscular or subcutaneous injection. All of these infants have very limited muscle mass, and the hypertonicity or pH of some drugs can cause severe pain which then causes anorexia or local necrosis of muscle.

Extended courses of antibiotic therapy often cause severe disturbances of the normal intestinal flora, and protracted diarrhea can result. For this reason, following antibiotic therapy we feed reconstituted, lyophilized cultures of E. coli, Enterobacter spp, Enterococcus spp, and Lactobacillus acidophilus.

We have observed a syndrome that includes hematuria, bloating, dehydration, convulsions, and death, and seems to be due to accidental feeding of hyperosmotic formula. This condition has been well documented in human nursery practice, and the basic pathologic mechanism involves severe electrolyte shifts across the gastric mucosa. The formula we were feeding at the time the most severe cases developed had a concentration of 1,039 milliosmols/kg. Normal marmoset milk is 292 milliosmols/kg. Changing to a formula with concentration similar to natural milk alleviated the condition.

At present, hand-rearing New World primate infants is still rudimentary. Survival rates for infants that are healthy at birth to 1 year of age do not yet exceed 50%. With added experience additional basic physiological data and mechanisms of disease should be more precisely defined, resulting in greater survival and usefulness of these unique primate species.

ACKNOWLEDGMENTS

This work was supported by NIH contract NO1 CP 6-1006 from the National Cancer Institute. This work was conducted

according to the principles of animal care promulgated by the
NAS-NRC.   (See "Guide for Laboratory Animal Facilities and
Care," DHEW Publication No. (NIH) 73-23.)

REFERENCES

Hampton, S. H. and Hampton, Jr., J. K. Rearing marmosets from
    birth by artificial laboratory techniques. Lab. anim. Care
    17:1-10, 1967.
Valerio, D. A., Darrow II, C. C., and Martin, D. P. Rearing of
    infant simians in modified germfree isolators for oncogenic
    studies. Lab. anim. Care 20:713-719, 1970.
Wolfe, L. G., Odgen, J. D., Deinhardt, J. B., Fisher, L., and
    Deinhardt, F. Breeding and hand rearing marmosets for viral
    oncogenesis studies. Pp. 145-147 in: Primate Breeding, W. I.
    B. Beveridge (Ed.), Basel: Karger, 1972.

CHAPTER 24

HAND-REARING Saguinus AND Callithrix
GENERA OF MARMOSETS

J. D. Ogden

Rush-Presbyterian-St. Lukes Medical Center
Chicago, Illinois

INTRODUCTION

Marmosets and tamarins, small monkeys native to Central
and South America (Hershkovitz, 1966), have been used extensively
in biomedical research during the past decade (Wolfe and
Deinhardt, 1972; Deinhardt, 1973; Falk, 1974; Wolfe, Deinhardt,
Ogden, Adams, and Fisher, 1975). In 1965, a nursery for hand-
rearing neonatal marmosets was established in our laboratory, in
conjunction with establishment of a breeding colony of marmosets
for production of neonates for viral oncology studies. The nursery
was required for rearing parentally neglected, uninoculated animals
and animals inoculated with oncogenic viruses. Neonates were
used in the experimental studies because of their greater susceptibility to selected oncogenic viruses and, once inoculated, were
reared in isolation from the breeding colony to avoid contamination
of the breeders and to permit closer monitoring for experimentally
induced disease. Subsequently, hand-reared animals also were used
in our laboratory in studies of slow virus diseases of the central
nervous system.

The initial results of hand-rearing were not particularly
successful. The small size of newborns (30-50 g) posed a problem
in handling and feeding, and their nutritional requirements were
unknown. Infants often succumbed early in life with aspiration
pneumonias and/or enteritis, and those that survived frequently
developed osteoporosis. Because of improved husbandry and discovery of the Vitamin D3 requirement of marmosets, these problems are seldom encountered today and the survival rate of hand-
reared animals is equivalent to that of parent-reared animals. Our

313

average nursery population currently numbers approximately 24 animals, cared for by four attendants who work two 8-hr shifts (overlapping) from 6:00 a.m. to 8:30 p.m.

HOUSING

Offspring destined for hand-rearing are usually separated from their parents within the first 24-48 hr after parturition and transferred to the nursery. The infant marmosets are maintained in standard isolette incubators (Air Shields, Inc.) equipped with two open-ended enclosures made of 3/4-inch mesh galvanized wire cloth. One infant is housed per enclosure, permitting two infants to use an incubator at once. Separation with the wire enclosures is necessary because a stronger infant will "ride" a weaker one, sometimes resulting in death of the weaker animal from exhaustion. The riding tendency apparently disappears at about 4 months of age. Each animal unit or enclosure contains a cloth surrogate composed of 2-inch tubular stockinette packed with cotton and suspended at a 60° angle from the top of the enclosure by a rubber band. Soiled surrogates are discarded and replaced. The floor of each unit is covered with a human infant disposable diaper which is replaced 1-2 times per shift. When the animal reaches 4-6 weeks of age (natural weaning time) the diaper is removed, permitting the infant to crawl and walk on the galvanized wire (infants are usually crawling up and down the sides of the wire enclosure by this time), and the cloth surrogate is replaced by a 1-inch transverse wooden perch. A layer of absorbent paper with a moisture-resistant backing is placed beneath each wire unit to facilitate absorption of urine, feces, and drops of formula and water.

When infants, which initially weigh 30-50 g, reach a morning body weight of 150 g, they are removed from the incubators and transferred to modified rabbit cages in rooms designed for juveniles and young adults. Experimental animal are segregated into rooms containing other animals previously injected with the same or similar inoculum. To guard against infection, caretakers work first in the uninoculated rooms, then change shoe covers, gown, face mask, and gloves. This same procedure is then followed in the inoculated rooms. Rubber gloves are changed after each animal is handled in all rooms. Care of the young animals remains the responsibility of the nursery staff until the animals reach 12-18 months of age, whereupon they are transferred to general holding rooms and, if not previously designated, are assigned to either an experimental group or the breeding colony.

AMBIENT CONDITIONS

The incubators are maintained at 90° F and 70-80% relative humidity (R.H.) for 4 weeks and then the temperature and R.H. are gradually reduced to room conditions of 80° F and 50% R.H. by the time the infants are transferred from the incubators to modified rabbit cages at 10-12 weeks of age. The latter conditions are standard for all juvenile and young adult animals whereas for the more mature, well-acclimated adults, the temperature is maintained at about 74° F.

DIET AND FEEDING REGIMEN

The routine currently employed (Table 1) evolved from experience gained over the past 12 years although deviations from the routine are sometimes necessary because of the size or species of the animal.

The neonate is fasted for 4-6 hr after separation from the parents. The first feeding generally consists of 0.50 ml of 10% dextrose plus 1 drop of pediatric ampicillin (100 mg/ml) followed by three feedings (0.50 ml) of a 1:1 mixture of 10% dextrose and SMA-S-26 powder (Wyeth Labs) with Sustagen (Meade-Johnson Co.). Then 0.50 ml of undiluted SMA-S-26 supplemented with Sustagen is offered every 3 hr for the next 2-3 days at which time the volume is increased by 0.25 ml; the volume subsequently is increased by 0.25-ml increments as the animal's weight and condition indicate. By 2 weeks of age, the ampicillin therapy is terminated and the feeding interval is extended to every 4 hr and the infants are partially self-feeding; baby cereal is introduced, a pediatric vitamin preparation (Vi-Daylin, Ross Laboratories) is offered once daily, and ampicillin is discontinued. By 4-6 weeks of age the neonate is totally self-feeding and feeding intervals are increased to 6 hr; 1 drop of aqueous vitamin D3 (10,000 i.u./cc) is added to the formula at each feeding. At 8 weeks of age, formula, fresh fruit, chopped canned marmoset diet, and powdered monkey blocks are offered twice daily. This regimen is maintained until the animals are 14 weeks of age (2 weeks after transfer from the isolettes to the modified rabbit cages) at which time they are offered canned marmoset diet and monkey blocks and tap water ad libitum; fresh fruit and pediatric vitamin supplement are offered on alternate days.

Small vaccine bottles (3 ml) fitted with nipples are used for hand-feeding. We manufacture the nipples from a commercially available latex (Lotol, Uniroyal Inc., Chicago, IL), and use a

TABLE 1. HAND-REARED MARMOSET DIETARY REGIMEN

Age (Days)	Diet	Volume (cc)	Frequency
1	10% Dextrose + 1 drop Ampicillin (100 mg/ml) (hand-fed, drop by drop)	0.50	once
1	½ 10% Dextrose, ½ SMA with Sustagen + 1 drop Ampicillin (hand-fed, drop by drop)	0.50	Q3H
2-4	SMA with Sustagen + 1 drop Ampicillin (hand-fed, drop by drop)	0.50	Q3H
5	SMA with Sustagen + 1 drop Ampicillin (hand-fed, sucking)	0.75	Q3H
7	SMA with Sustagen + 1 drop Ampicillin (hand-fed, sucking)	1.00	Q3H
9	SMA with Sustagen + 1 drop Ampicillin (hand-fed, sucking)	1.25	--
11	SMA with Sustagen + 1 drop Ampicillin (hand-fed, sucking)	1.50	Q3H
14-17	SMA with Sustagen + 1.0cc baby cereal (start self-feeding)	2.00-2.50	Q4H B.I.D.
21-28	SMA with Sustagen + 2.0cc baby cereal + pediatric vitamin + tap water (self-feeding)	3.00	Q4H B.I.D. ad lib ad lib
28-42	SMA with Sustagen + 2.0cc baby cereal + powdered monkey blocks + pediatric vitamins + tap water (self-feeding)	3.50-5.0	Q4H B.I.D. ad lib ad lib ad lib

Age (Days)	Diet	Volume (cc)	Frequency
42-56	SMA with Sustagen	7-10	Q6H
	+ canned diet		U.I.D.
	+ fresh fruit		U.I.D.
	+ cracked blocks		ad lib
	+ pediatric vitamins		ad lib
	+ Vitamin D_3		U.I.D.
	+ tap water		ad lib
56-84	SMA with Sustagen	15-40	B.I.D.
	+ canned diet		U.I.D.
	+ fresh fruit		U.I.D.
	+ whole blocks		ad lib
	+ pediatric vitamins		ad lib
	+Vitamin D_3		U.I.D.
	+ tap water		ad lib
84-98	SMA with Sustagen	40	B.I.D.
	+ canned diet		U.I.D.
	+ fresh fruit		U.I.D.
	+ whole blocks		ad lib
	+ pediatric vitamins		ad lib
	+ Vitamin D_3		U.I.D.
	+ tap water		ad lib
98-140	Canned diet	--	U.I.D.
	+ fresh fruit 3x/week		
	Pediatric vitamins 3x/week		
	Tap water		ad lib

Q3H = every 3 hours   Q4H = every 4 hours
Q6H = every 6 hours   B.I.D. = twice a day
U.I.D. = once a day   ad lib = available at all times

SMA: Wyeth Company, Philadelphia, PA
Sustagen: Mead-Johnson Company, Evansville, ID
Vi-Daylin: Ross Laboratories, Columbus, OH

molding form consisting of a 3-ml disposable plastic syringe barrel with an attached 1.5 x 22 gauge needle; the needle is simply cut off at the desired length (usually about 1.0 cm long). During the initial feedings, formula is fed by drops through the nipples, with the feeding of 0.50 ml taking approximately 10 minutes. Patient, conscientious nursery technicians are essential for successful hand-rearing.

Animals are introduced to self-feeding by placing formula in 10-ml Wasserman tubes and fastening the tubes to the wire enclosures. The tubes are fitted with a stainless steel sipper tube which is guarded with a rubber mouthpiece to minimize trauma to the lips and tongue. Once the animals consume more than 10 ml of formula per feeding, the formula is offered in a 4-oz French square bottle equipped with sipper tube.

Interspecies variations in feeding techniques are notable but not pronounced. Callithrix jacchus jacchus infants, which are among the lowest in birth weights, seem to accept our nipples and formulas more readily and become self-feeding at an earlier age than either Saguinus fuscicollis or S. oedipus oedipus infants, which are slowest in establishing a sucking reflex. All species are self-feeding by 2-3 weeks of age.

MONITORING OF BODY WEIGHT

The newborns are weighed twice daily, before the first feeding and after the last one, for the first 3 days and then in the morning 3 days per week until they reach 50 g, whereupon they are weighed weekly. The animals are placed in a cylindrical, 1-quart waxed cardboard container (separate container for each animal) and weighed on a Mettler electric scale. Upon transfer to rabbit cages at 10-12 weeks of age, the animals are weighed weekly for 2 weeks, monthly for 3 months, and then every 6 months, using a triple-beam balance (O'Haus) and a perforated aluminum basket. Any deviation from the anticipated weight gain results in a closer scrutiny of the animal, appropriate modification of husbandry procedures, and more frequent (weekly) weighings until the animal's condition returns to normal.

30-DAY SURVIVAL RATE

During the 3-year period 1974-1976, we hand-reared 228 "normal" marmoset infants of which 197 (86.4%) survived longer than 30 days. In addition, we placed 116 parentally neglected newborns in the nursery of which 38 (32.9%) survived 30 days or longer. Overall, the 30-day survival rate of 68.3% for the 344 marmosets hand-reared during this period compared favorably with the 63.2% rate for parent-reared offspring. We use the 30-day survival period as a measure for comparing hand-rearing with parent-rearing because in our experience fatalities after 30 days of age are very rarely related to the rearing method. There was no appreciable difference noted in survival rates between any of the various species.

Body weights at 2 years of age were some 10% greater in hand-reared than in parent-reared offspring (Wolfe, Ogden, Deinhardt, Fisher, and Deinhardt, 1972).

REFERENCES

Deinhardt, F. Herpesvirus saimiri. Pp. 595-625 in: The Herpes-viruses, New York: Academic Press, 1973.

Falk, L. A. Oncogenic DNA viruses of nonhuman primates: A review. Lab. anim. Sci. 24:182-192, 1974.

Hershkovitz, P. Taxonomic notes on tamarins, genus Saguinus (Callithricidae, primates), with descriptions of four new forms. Folia Primat. 4:381-395, 1966.

Wolfe, L. G. and Deinhardt, F. Oncornaviruses associated with spontaneous and experimentally induced neoplasia in non-human primates. Pp. 176-196 in: Medical Primatology 1972, Basel: Karger, 1972.

Wolfe, L. G., Deinhardt, F., Ogden, J. D., Adams, M. R., and Fisher, L. E. Reproduction of wild-caught and laboratory-born marmoset species used in biomedical research (Saguinus sp., Callithrix jacchus). Lab. anim. Sci. 25:802-813, 1975.

Wolfe, L. G., Ogden, J. D., Deinhardt, J. B., Fisher, L., and Deinhardt, F. Breeding and hand-rearing marmosets for viral oncogenesis studies. Pp. 145-157 in: Breeding Primates, Basel: Karger, 1972.

CHAPTER 25

WEIGHT GAINS AND SEQUENCE OF DENTAL ERUPTIONS

IN INFANT OWL MONKEYS (Aotus trivirgatus)

R. D. Hall, R. J. Beattie, and G. H. Wyckoff, Jr.

Division of Veterinary Resources
Walter Reed Army Instiutue of Research
Walter Reed Army Medical Center
Washington, DC

INTRODUCTION

Data on the growth and development of infant owl monkeys (Aotus trivirgatus) are minimal. The average birth weight and body lengths of 13 newborn owl monkeys have been published (Elliott, Sehgal, and Chalifoux, 1976), but data on weight gains of infant owl monkeys are not available. Monitoring body weight is, in our experience, the simplest method for evaluating the health and development of infant Aotus. Failure to make expected weight gains during maternal dependence is particularly critical, and may indicate that medical treatment of infant or maternal disease is necessary.

The sequence of eruption of permanent teeth in Aotus has been studied by examining the skulls of immature monkeys (Della Serra, 1953; Thorington and Vorek, 1976). Since the subjects examined were of unknown ages, the ages at which permanent teeth usually erupt could not be determined. The sequence and time of eruption of deciduous and permanent teeth have not been reported. Precise information is required on both sequence and time of eruption of deciduous and permanent teeth if these criteria are to be used to estimate the age of immature monkeys.

Recent births in our Aotus colony have afforded us the opportunity to study infant weight gains and dental eruptions. These data are presented to assist others in monitoring the growth and development of infant owl monkeys and to provide a basis for estimating the age of immature owl monkeys.

MATERIALS AND METHODS

The infant Aotus examined were born in the breeding colony at the Walter Reed Army Institute of Research (WRAIR). The colony and the husbandry techniques employed have been described elsewhere (Hall, Renquist, Montrey, Beattie, and Wyckoff, 1977). Eight infants ranged in age from 7.5 to 17.5 months when the study began, and others were born at various times during the ensuing 48 months. Siblings remained with their parents until they were 6 months of age or older unless their separation was necessary for survival.

Parents of the newborn were chemically immobilized with ketamine HCl (Vetalar, Parke Davis and Co., Detroit, MI) 1-3 days after parturition, and infants and parents were individually weighed and examined. Examination of the parents after parturition was necessary to assess their general health and potential for rearing their infant. Furthermore, handling the parents at that time reduced the tendency for the parents to neglect their infant because of human manipulation. For the second examination only the parent carrying the sibling was chemically restrained. Chemical restraint usually was not necessary for subsequent examinations.

Body weights were obtained and dental examinations performed twice weekly on Monday and Friday until all deciduous teeth had erupted, and then biweekly on Friday until all permanent teeth had erupted. Eruption was recorded when any part of the tooth crown was observed penetrating the gingiva. Body weights were obtained on a direct reading counter scale (Model 4030, Toledo Scale Co., Toledo, OH) and results were recorded to the nearest gram. A chart of weight gains was plotted for each monkey using available data points. The mean and median body weights of available monkeys were calculated for weekly intervals.

RESULTS

The body weights of 36 apparently normal 1- to 3-day-old owl monkeys ranged from 69 to 114 g with an average of 90 g $\pm$ 11.2 g ($\pm$1 S.D.). Although the average weight for the 23 males (88 g $\pm$ 11.0 g) was less than the weight of the 13 females (94 g $\pm$10.9 g), the difference was not significant at the 0.1 level. Five additional monkeys were clinically premature at birth and had an average body weight of 63 g $\pm$ 9.5 g (range 55-77 g). All died within 5 days of birth.

Data on body weights during growth were available for 24 monkeys through the 6th week of age, 19 monkeys through the 13th week, 14 monkeys through the 26th week, and 8 monkeys through the 39th week. A representative weight curve for an individual monkey through its first 12 weeks of age is shown in Figure 1. Body weights usually showed a steady increase through the first 4-6 weeks. Then, in some monkeys there was a 4- to 7-day interruption in weight gain or a transient weight loss between the 4th and 8th weeks. This occurred soon after eruption of the last deciduous teeth and may be related to weaning. We have observed the same pattern of weight change when hand-reared infants were weaned from infant formula to semisolid or solid food at approximately 6 weeks of age.

The weekly mean and median values for body weights of Aotus calculated from birth to 9 months of age were within 3% of each other for 39 of 40 observations. Weights were recorded for only three monkeys between the 9th and 12th months of age. A greater variation was seen between these mean and median values because one monkey was considerably heavier than the others. A weight chart for Aotus was plotted using median weights (Figure 2). Percentile confidence lines were constructed so that the body

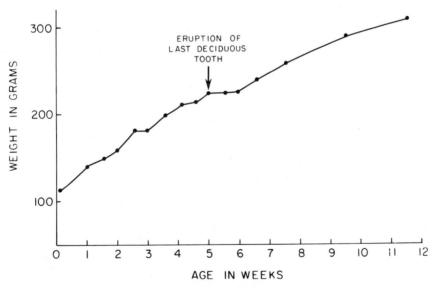

Figure 1. Representative weight gain chart for individual owl monkeys.

weights of 90% of the monkeys observed were below the upper (90 percentile) line and the weights of 10% of the monkeys were below the lower (10 percentile) line.

The rate of weight gain was approximately 16 g per week from birth through the 15th week of age and 10 g per week thereafter through the 52nd week (Figure 2). Weight gains were considered most critical during the initial 4 weeks when the infants were maternally dependent. Among the 24 Aotus examined during this period, weight gains ranged from 9 to 26 g per week with an average gain of 18 g per week. The growth rate for most individual monkeys was almost linear through the first 3 months of age. Thereafter, greater fluctuations in individual weighings occurred, but the general weight gain pattern for each monkey was consistent with the pattern described for the group.

The age at eruption of the deciduous teeth was observed and recorded for 28 infants (Table 1). Owl monkeys are born without teeth. Eruptions usually began with the central incisors (I_1) at 1-3 weeks of age and were completed at 4-7 weeks

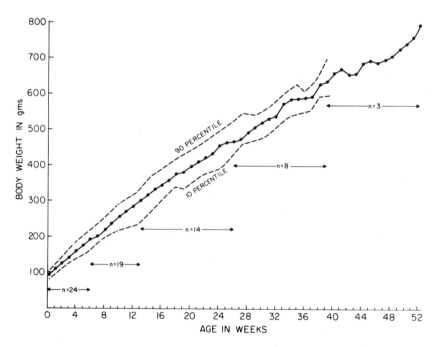

Figure 2. Median weight gain chart for a group of owl monkeys.

of age when the posterior premolars (P_3) emerged. A general eruption sequence of I_1, I_2, P_1, P_2, C and P_3 was observed; however, 42% of the animals had minor variations in this sequence. P_1 and P_2 or I_2 and P_1 frequently erupted simultaneously. The eruption of mandibular I_2 incisors preceded the eruption of the ipsilateral maxillary I_2 incisors on 39 of 56 (70%) occasions; both erupted concurrently on the other 17 occasions. Otherwise, there was no consistency in maxillary or mandibular teeth erupting first.

TABLE 1. MEAN ERUPTION TIME FOR DECIDUOUS TEETH OF 28 OWL MONKEYS

Tooth		Mean	Age in Weeks 1 Standard Deviation	Range	% of Times Observed Erupting First*
I_1	MX	2.3	0.7	1.0 – 3.5	26
	MD	2.3	0.8	0.5 – 3.5	26
I_2	MX	3.5	0.9	1.5 – 4.5	0
	MD	2.9	0.8	1.0 – 4.5	70
C	MX	4.7	1.0	1.5 – 6.0	27
	MD	4.6	1.2	2.0 – 6.5	23
P_1	MX	3.5	0.7	1.5 – 5.0	13
	MD	3.4	0.7	2.0 – 5.0	27
P_2	MX	3.8	0.8	2.0 – 5.5	21
	MD	3.8	0.9	2.0 – 5.0	18
P_3	MX	5.6	1.2	3.0 – 8.0	13
	MD	5.5	1.1	3.0 – 7.5	27

*Percentage of observations in which the maxillary or mandibular tooth erupted before its ipsilateral counterpart. The difference between the sum of MX + MD and 100% equals the percentage of observations in which the maxillary and mandibular teeth erupted together. MX = maxillary tooth, MD = mandibular tooth.

TABLE 2.   MEAN ERUPTION TIME FOR PERMANENT TEETH
            OF OWL MONKEYS

Tooth		No. of Observations*	Age(Months) Mean	Age(Months) 1 Standard Deviation	Age(Months) Range	% of Times Observed Erupting First**
I_1	MX	24	9.4	0.8	8.0 - 11.0	55
	MD	18	9.6	0.6	9.5 - 11.0	9
I_2	MX	13	10.8	0.6	10.0 - 12.0	64
	MD	8	10.6	0.4	10.0 - 11.0	0
C	MX	18	15.0	1.3	13.5 - 18.0	29
	MD	15	14.0	1.0	12.5 - 16.0	43
P_1	MX	12	12.4	1.1	11.0 - 14.5	33
	MD	13	12.4	0.9	11.5 - 14.0	17
P_2	MX	12	11.9	1.0	10.5 - 14.0	67
	MD	13	12.3	0.8	11.5 - 14.0	0
P_3	MX	12	11.5	0.9	10.0 - 13.0	13
	MD	16	11.0	0.7	10.0 - 12.0	50
M_1	MX	36	4.9	0.6	4.0 - 6.0	0
	MD	38	4.3	0.5	3.5 - 5.0	84
M_2	MX	28	7.3	0.6	6.5 - 8.5	0
	MD	32	6.4	0.8	5.0 - 7.5	100
M_3	MX	17	11.2	1.1	10.0 - 13.0	0
	MD	18	9.9	1.2	8.5 - 11.5	91

*Observation is the eruption of a single tooth.

**Percentage of observations in which the maxillary or mandibular
tooth erupted before its ipsilateral counterpart. The difference
between the sum of MX + MD and 100% equals the percentage of
observations in which the maxillary and mandibular teeth erupted
togethers. MX = maxillary tooth, MD = mandibular tooth.

The average ages at which permanent teeth erupted are shown in Table 2. Each observation represents the eruption of a single tooth. The anterior molars (M_1) were the first permanent teeth to erupt at 3.5-5 months of age. The eruption sequence of permanent teeth was M_1, M_2, I_1, M_3, I_2, P_3, P_2, P_1, and C for mandibular teeth and M_1, M_2, I_1, I_2, M_3, P_3, P_2, P_1, and C for maxillary teeth. Mandibular molars (M_1, M_2, and M_3) usually erupted before their maxillary counterparts while maxillary incisors (I_1 and I_2) and middle premolars (P_2) usually erupted before their mandibular counterparts. In 8 of 9 monkeys observed, all permanent teeth had erupted by 14.5-16 months of age.

DISCUSSION

The average weight of 90 g for the 36 infants weighed at 1-3 days of age was similar to the average birth weight of 92 g previously reported for Aotus (Elliott et al., 1976). Aotus weighing less than 75 g at birth experience a higher rate of neonatal mortality than do heavier newborns and require close monitoring. Light-weight monkeys, especially those born prematurely, are more likely to be too weak to clasp the fur of their parents and will die from hypothermia within 1-2 hr if neglected.

The birth weight alone is less valuable for establishing the health of a newborn than is monitoring its rate of weight gain. Our data show that Aotus should gain at least 10 g per week for the first 4-6 weeks or until all deciduous teeth have erupted. Failure to achieve expected weight gains may indicate poor health of the infant or its mother, agalactia of the mother, or simply parental neglect. Weight loss is most critical during maternal dependence, and hand-rearing may be necessary if weight loss occurs. In our experience, infants that were hand-reared because of agalactia or parental neglect have gained weight at a rate comparable to that of parent-reared animals. Greater variation among individual weighings is expected after weaning, but monitoring should be continued to ensure expected trends in weight gains.

The observed sequence of eruption of permanent teeth generally agreed with the sequence previously reported for Aotus (Della Serra, 1953; Thorington et al., 1976), and is similar to the sequence reported for squirrel monkeys (Long and Cooper, 1968). Our data on the age of Aotus at eruption of their permanent and deciduous teeth will now permit relatively accurate estimates of ages for immature owl monkeys on the basis of dental eruptions.

ACKNOWLEDGMENTS

In conducting the research described in this report, we adhered to the "Guide for the Care and Use of Laboratory Animals," as promulgated by the Institute of Laboratory Animal Resources, National Academy of Sciences, National Research Council. We wish to thank Mrs. Ann Abston and Mr. Pat Phillips for their technical assistance on this project, and Drs. E. H. Stephenson and T. J. Keefe for their editorial comments.

REFERENCES

Della Serra, O.   A sequencia eruptiva dos dentes definitivos nos simios Platyrrhina e sua interpretacao filogenetica.   Folia clin. biol. 19:41–46, 1953.

Elliott, M. W., Sehgal, P. K., and Chalifoux, L. V. Management and breeding of Aotus trivirgatus. Lab. anim. Sci. 26:1037–1040, 1976.

Hall, R. D., Renquist, D. M., Montrey, R. D., Beattie, R. J., and Wyckoff, G. H., Jr. Management of an experimental breeding colony of Aotus monkeys. Submitted for publication, 1977.

Long, J. O. and Cooper, R. W.   Physical growth and dental eruption in captive-bred squirrel monkeys, Saimiri sciureus (Leticia, Colombia). In: The Squirrel Monkey, L. A. Rosenblum and R. W. Cooper (Eds.), New York: Academic Press, 1968.

Thorington, Jr., R. W. and Vorek, R. E.   Observations on the geographic variation and skeletal development of Aotus. Lab. anim. Sci. 26:1006–1021, 1976.

INDEX

Age assessment: See Physical growth; Skeletal maturation

Albumin: levels in neonatal Macaca nemestrina, 121-22

Anesthesia: for prepartum examination, 4; for amniocentesis, 4; use of ketamine hydrochloride, 4, 5, 28, 186, 300; for Caesarean section deliveries, 5, 6; use of halothane, 6; for postpartum manipulation, 28; for Papio, 146; for Hylobates, 300; for Aotus, 322

Anthropometrics: See Physical growth

Antibiotics: use in Hylobates, 292-95, 304-05; use in Cebus, Aotus, and Callithrichids, 309-11, 315-16

Aotus: birth weight statistics, 308, 322, 327; feeding regimen, 308, 311; clinical management, 309-11; weight gain, 310, 323-24, 327; anesthesia for, 322; dental development, 324-27

Apnea: normal respiration in neonatal Macaca nemestrina, 134-40, 217-22; therapy for, 177-78; monitor for, 203-12; central vs. peripheral, 223

Baboon: See Papio

Bilirubin: levels in normal-weight vs. low-birth-weight Macaca nemestrina, 119-21; hyperbilirubinemia, 122

Birth weight: Macaca mulatta, 74, 168-71; Macaca nemestrina, 81, 169, 190-91; to determine gestational age, 108; Papio, 151; Macaca fascicularis, 168, 170, 174; effects of caging on, 168, 170; Hylobates, 295, 300; Callitrichids, 307-08, 313; Cebus, 308; Aotus, 308, 322, 327

Blood typing: for materno-fetal compatibility, 39-40; for blood transfusion, 40

Caesarean section: technique, 5-6; factors in determining necessity for, 7-8, 14-17; effect on physiological development of infant, 134-40 passim; effect on weight gain in infant, 198-99

Clinical problems: erythroblastosis fetalis, 35-40 passim; in Saimiri, 158-59; general management, 183-84; diarrhea, 183-84; heat therapy, 184; survey of diseases in Macaca nemestrina, 227-35 passim; infection in Aotus, Callitrichids, and Cebus, 309-11

Callitrichids: birth weight statistics, 307-08, 313; clinical management, 309-11; surrogate mother for, 314; environment for, 314-15; antibiotics for, 315; feeding regimen, 308, 311,